WHITE NOISE ANALYSIS AND QUANTUM INFORMATION

LECTURE NOTES SERIES
Institute for Mathematical Sciences, National University of Singapore

Series Editors: Chitat Chong and Kwok Pui Choi
Institute for Mathematical Sciences
National University of Singapore

ISSN: 1793-0758

*For the complete list of titles in this series, please go to
http://www.worldscientific.com/series/LNIMSNUS

Lecture Notes Series, Institute for Mathematical Sciences, National University of Singapore

Vol. 34

WHITE NOISE ANALYSIS AND QUANTUM INFORMATION

Editors

Luigi Accardi
Università di Roma Tor Vergata, Italy

Louis H Y Chen
National University of Singapore, Singapore

Takeyuki Hida
Nagoya University and Meijo University, Japan

Masanori Ohya
Tokyo University of Science, Japan

Si Si
Aichi Prefectural University, Japan
Yangon University, Myanmar

Noboru Watanabe
Tokyo University of Science, Japan

World Scientific

EW JERSEY · LONDON · SINGAPORE · BEIJING · SHANGHAI · HONG KONG · TAIPEI · CHENNAI · TOKYO

Published by

World Scientific Publishing Co. Pte. Ltd.

5 Toh Tuck Link, Singapore 596224

USA office: 27 Warren Street, Suite 401-402, Hackensack, NJ 07601

UK office: 57 Shelton Street, Covent Garden, London WC2H 9HE

Library of Congress Cataloging-in-Publication Data

Names: Accardi, L. (Luigi), 1947– editor.

Title: White noise analysis and quantum information / edited by Luigi Accardi
  (University of Roma II, Tor Vergata, Italy), Louis H. Y. Chen (NUS, Singapore),
  Takeyuki Hida (Kyoto University, Japan), Masanori Ohya (Tokyo University of Science, Japan),
  Si Si (Aichi Prefectural University, Japan), Noboru Watanabe (Tokyo University of Science, Japan).

Description: New Jersey : World Scientific, 2017. | Series: Lecture notes series,
  Institute for Mathematical Sciences, National University of Singapore ; volume 34 |
  Includes bibliographical references and index.

Identifiers: LCCN 2017019449 | ISBN 9789813225459 (hardcover : alk. paper)

Subjects: LCSH: White noise theory. | Stochastic processes. | Spectral energy distribution.

Classification: LCC QA274.29 W45 2017 | DDC 519.2/2--dc23

LC record available at https://lccn.loc.gov/2017019449

British Library Cataloguing-in-Publication Data

A catalogue record for this book is available from the British Library.

Dedicated to Professor Takeyuki Hida

CONTENTS

FOREWORD

The Institute for Mathematical Sciences (IMS) at the National University of Singapore was established on 1 July 2000. Its mission is to foster mathematical research, both fundamental and multidisciplinary, particularly research that links mathematics to other efforts of human endeavor, and to nurture the growth of mathematical talent and expertise in research scientists, as well as to serve as a platform for research interaction between scientists in Singapore and the international scientific community.

The Institute organizes thematic programs of longer duration and mathematical activities including workshops and public lectures. The program or workshop themes are selected from among areas at the forefront of current research in the mathematical sciences and their applications.

Each volume of the *IMS Lecture Notes Series* is a compendium of papers based on lectures or tutorials delivered at a program/workshop. It brings to the international research community original results or expository articles on a subject of current interest. These volumes also serve as a record of activities that took place at the IMS.

We hope that through the regular publication of these *Lecture Notes* the Institute will achieve, in part, its objective of reaching out to the community of scholars in the promotion of research in the mathematical sciences.

April 2017

Chitat Chong
Kwok Pui Choi
Series Editors

PREFACE

According to the research program on Infinite Dimensional Analysis and Quantum Probability and their Applications (IDAQP), the workshop was held at the Institute for Mathematical Sciences (IMS) at the National University of Singapore (NUS) from 3rd of March to 7th of March, 2014. It was very nice to see that many eminent mathematicians have contributed mainly in the areas of White Noise Analysis, Quantum Probability and Quantum Information Theory and cultivated new ideas.

We believe that the research program is successful since the workshop meets the following goals of the program;

(1) making a bridge among the fields of infinite dimensional analysis and quantum probability,

(ii) reviewing the recent developments and the research on these interdisciplinary fields, mainly focused on quantum information theory and white noise analysis in line with IDAQP,

(iii) accelerating the interaction of infinite dimensional analysis and quantum probability, and promoting the highly interdisciplinary research on new exciting areas in mathematical sciences such as quantum information theory and white noise analysis.

The current volume collects three expanded lectures; Quantum Information, White Noise Analysis, and other fields on IDAQP.

We are very much grateful to the other members of the organizing committee such as

Masanori Ohya (Chief Organizer), Tokyo University of Science,
Louis Chen, National University of Singapore,
Takeyuki Hida, Nagoya University and Meijo University,
Luigi Accardi, Università di Roma Tor Vergata
for their invaluable advice.

Thanks are also due to all the participants of this program for their support and stimulating interactions during the workshop!

We would like to take this opportunity to thank Professor Louis Chen, Former Director of IMS, for his leadership in creating an exciting environment for mathematical research in IMS and for his guidance throughout our program.

The expertise and dedication of all IMS staff contributed essentially to the success of this program. We would like to acknowledge IMS for providing financial support to the program.

January 2017

Luigi Accardi
Università di Roma Tor Vergata

Louis H. Y. Chen
National University of Singapore

Takeyuki Hida
Nagoya University
Meijo University

Masanori Ohya
Tokyo University of Science

Si Si
Aichi Prefectural University
Yangon University

Noboru Watanabe
Tokyo University of Science

Volume Editors

EXTENSIONS OF QUANTUM THEORY CANONICALLY ASSOCIATED TO CLASSICAL PROBABILITY MEASURES

Luigi Accardi

Centro Vito Volterra, Università di Roma Tor Vergata, Italy
http://volterra.mat.uniroma2.it
accardi@volterra.mat.uniroma2.it

1. Introduction

For several decades the mathematical model of Quantum Probability (QP) has been considered as a generalization of classical probability. However some discoveries of the past 15 years [1, 6, 7, 3, 4, 5], show that the whole quantum theory, including quantum fields, is not a generalization, but rather a **deeper level** of classical probability. The fecundity of these new ideas is testified by the fact that they have allowed to solve a multiplicity of long-standing open problems both in classical probability and in the theory of orthogonal polynomials. In particular, combining classical probability with the theory of orthogonal polynomials in 1 or several real variables, it is possible to prove that the canonical commutation relations, both Fermi and Bose (and in fact even their q-deformations), arise canonically from the Bernoulli and Gaussian random variables respectively.

More generally one can prove that there is a one-to-one correspondence between Heisenberg-type commutation relations and equivalence classes of probability measures on $\mathbb{R}$ with all moments. The equivalence relation being defined, in the one-dimensional case, by the fact that all measures in a class share the same principal Jacobi sequence.

To each of these equivalence classes it is canonically associated a *free evolution*, generalizing the classical harmonic oscillator evolution.

The characterization of the equilibrium states with respect to any such evolution naturally leads to a generalization of the Planck's factor. Similar arguments, applied to the recently introduced *local equilibrium states*, lead to non-linear extensions of the Planck's factor and non-linear Gibbs states.

Being functorial, the above construction also provides a generalization of the second quantization procedure both at Hilbert space (Fock) and *-algebra level and in some special cases (e.g. probability measures on $\mathbb{R}^d$ with compact support) even at C^*-algebra level. However in general the class of morphisms will be much narrower than in usual second quantization.

This fact supports the intuition that the new quantizations have a physical meaning in terms of non-linear completely integrable classical systems.

In this chapter, the discussion is limited to the 1-dimensional case, i.e. to measures on $\mathbb{R}$ rather than on $\mathbb{R}^d$ and, after recalling, in Section 2, the notion of *quantum decomposition of a classical random variable* (with all moments) we deduce, at the end of Section 4, the commutation relations among the associated *creation, annihilation and preservation* (CAP) operators and we show, in Section 4.3, how the restriction to the Gaussian or Poisson case gives rise to the usual Heisenberg commutation relations.

The table at the end of Section 4.3 lists the commutation relations, orthogonal polynomials and their generating functions for some well-known classes of probability measures.

In Section 5 we further restrict to polynomially symmetric probability measures and for them, in Section 5.1, we construct the free evolutions, generalizing the usual quantum harmonic oscillator, and, in Section 5.2, the associated equilibrium states or weights.

2. The *-Algebra of Polynomial Functions

$$\mathcal{P} := \mathcal{P}(\mathbb{R}) := \text{ polynomial functions } \mathbb{R} \to \mathbb{C}$$

$$Q : x \in \mathbb{R} \to \sum_{n=0}^{N} a_n x^n \in \mathbb{C} \quad ; \quad a_n \in \mathbb{C}.$$

$\mathcal{P}$ is an abelian $*$-algebra for the pointwise operations with identity and involution given by complex coniugation

$$\left(\sum a_n x^n\right)^* = \sum \bar{a}_n x^n.$$

We denote

$$\mathcal{P}_{\mathbb{R}} \subset \mathcal{P}$$

the sub-algebra of polynomial functions with real coefficients.

2.1. *The polynomial filtration*

$\forall n \in \mathbb{N}$ define

$$\mathcal{P}_{0]} =: \mathcal{P}_0 = \mathbb{C} \cdot 1_{\mathcal{P}} \tag{1}$$

$$\mathcal{P}_{n]} := \{\text{Polynomials of degree } \leq n\}.$$

$$\mathcal{P}_{n],\mathbb{R}} := \mathcal{P}_{n]} \cap \mathcal{P}_{\mathbb{R}} = \{\text{Polynomials of degree } \leq n \text{ with real coefficients}\}.$$

Clearly

$$\mathcal{P}_{n]} \subset \mathcal{P}_{n+1]} \quad ; \quad \bigcup_{n \in \mathbb{N}} \mathcal{P}_{n]} = \mathcal{P}$$

and the same relations hold for the filtration $(\mathcal{P}_{N],\mathbb{R}})$.

2.2. *The monomial gradation*

$$x^n := \text{monomial of degree } n \; ; \; n \in \mathbb{N}$$

$$\mathcal{P} = \sum_{n \in \mathbb{N}}^{\cdot} \mathbb{C} \cdot x^n,$$

where $\cdot$ means direct sum of vector spaces.

2.3. *States on $\mathcal{P}$ and probability measures on $\mathbb{R}$*

Let

$$\mu \in Prob(\mathbb{R}).$$

Basic assumption: μ has moments of all orders:

$$\int_{\mathbb{R}} x^n \mu(dx) \text{ exists } \forall n \in \mathbb{N}.$$

Then the linear functional

$$\mu : Q \equiv \sum a_n x^n \to \mu(Q) := \sum a_n \int x^n \mu(dx)$$

is a state on $\mathcal{P}$:

$$\mu \geq 0 \qquad ; \qquad \mu(1) = 1.$$

Conversely, any state on $\mathcal{P}$ defines a probability measure on $\mathbb{R}$ (in general more than one). μ defines a **pre-scalar product** on $\mathcal{P}$:

$$\langle P, Q \rangle := \mu(P^*Q) = \int_{\mathbb{R}} \overline{P(X)} Q(x) \mu(dx). \qquad (2)$$

This is the restriction to $\mathcal{P}$ of the scalar product in $L^2(\mathbb{R}, \mu)$.

2.4. *Connection with random variables*

Fix a linear basis in $\mathbb{R}$

$$0 \neq e_1 \in \mathbb{R}$$

and define **the coordinate map**

$$X \equiv X_{e_1} \qquad : \qquad xe_1 \to x \in \mathbb{R}.$$

$(\mathbb{R}, \mu)$ is a classical probability space. Then X is a real valued random variable on $(\mathbb{R}, \mu)$. It is known that, up to stochastic equivalence, any real valued random variable can be realized in this way.

2.5. *The cyclic representation of* $(\mathcal{P}, \langle \cdot, \cdot \rangle)$

The pair $(\mathcal{P}, \langle \cdot, \cdot \rangle)$

$$\langle P, Q \rangle := \mu(P^*Q) = \int_{\mathbb{R}} \overline{P(X)} Q(x) \mu(dx) \qquad (3)$$

is a pre-Hilbert space. Denote

$$\Phi_0 := 1_{\mathcal{P}}$$

the identity of $\mathcal{P}$ considered as an element of $(\mathcal{P}, \langle \cdot, \cdot \rangle)$. When the pre-scalar product is given, we will use the notation

$$(\mathcal{P}, \langle \cdot, \cdot \rangle) := \mathcal{P} \cdot \Phi_0$$

i.e., when we want to emphasize that an element $Q \in \mathcal{P}$ is considered as a vector in the pre-Hilbert $(\mathcal{P}, \langle \cdot, \cdot \rangle)$, we write

$$Q \cdot \Phi_0 \qquad ; \qquad Q \in \mathcal{P}.$$

When considered as functions on $\mathbb{R}$, Q and $Q \cdot \Phi_0$ coincide.
For example $X^n \cdot \Phi_0 \in (\mathcal{P}, \langle \cdot, \cdot \rangle)$ denotes the n-th monomial function

$$(X^n \cdot \Phi_0)(x) := x^n \qquad ; \qquad x \in \mathbb{R} , \ n \in \mathbb{N}.$$

2.6. *Real valued random variables as symmetric operators*

The coordinate function X defines the multiplication operator (**position operator**) on $\mathcal{P}$:

$$(X(Q \cdot \Phi_0))(x) := xQ(x) \; ; \; x \in \mathbb{R} \, , Q \in \mathcal{P}.$$

This defines an action of $\mathcal{P}$ on the pre-Hilbert space

$$(\mathcal{P}, \langle \cdot, \cdot \rangle) \equiv \mathcal{P} \cdot \Phi_0$$

by (left) multiplication. $X : \mathcal{P} \to \mathcal{P}$ is a **symmetric linear operator on** $(\mathcal{P}, \langle \cdot, \cdot \rangle)$:

$$\langle Xf, g \rangle = \langle f, Xg \rangle.$$

Theorem: A pre-scalar product $\langle \cdot, \cdot \rangle$ on has the form

$$\langle P, Q \rangle := \mu(P^*Q) = \int_{\mathbb{R}} \overline{P(X)} Q(x) \mu(dx) \tag{4}$$

for some probability measure μ on $\mathbb{R}$ if and only if the position operator is symmetric with respect to $\langle \cdot, \cdot \rangle$.

We identify $\mathcal{P}$ to a sub-algebra of linear operators on $(\mathcal{P}, \langle \cdot, \cdot \rangle)$. Sometimes, when we want to emphasize that an element $Q \in \mathcal{P}$ is considered as an operator on $(\mathcal{P}, \langle \cdot, \cdot \rangle)$, we write simply Q or sometimes $Q(X)$.

3. The μ-Orthogonal Polynomials

Define

$$\Phi_0 := 1_{\mathcal{P}} \qquad ; \qquad \mathcal{P}_{0]} := \Phi_0 \Phi_0^*.$$

Suppose by induction to have constructed:

(i) an orthogonal system

$$\Phi_j \qquad ; \qquad j \in \{0, \dots, N-1\} \tag{5}$$

of linear generators of $\mathcal{P}_{N-1],\mathbb{R}} \cdot \Phi_0$ with the property that, for each monomial $j \in \{0, \dots, N-1\}$ either $\|\Phi_j\| = 1$ or $\|\Phi_j\| = 0$.

(ii) a sequence of orthogonal projections

$$P_{n]} : \xi \in \mathcal{P} \cdot \Phi_0 \to P_{n]}(\xi) := \sum_{j=0}^{n} \langle \Phi_j, \xi \rangle \Phi_j \; ; \quad n \in \{0, \dots, N-1\}.$$

Then one can define the vector

$$\Phi_N := \begin{cases} X^N \cdot \Phi_0 - P_{N-1]}(X^N \cdot \Phi_0) \ , \ \text{if} \ \|X^N \cdot \Phi_0 - P_{N-1]}(X^N \cdot \Phi_0)\| = 0 \\ \frac{X^N \cdot \Phi_0 - P_{N-1]}(X^N \cdot \Phi_0)}{\|X^N \cdot \Phi_0 - P_{N-1]}(X^N \cdot \Phi_0)\|} \ , \ \text{if} \ \|X^N \cdot \Phi_0 - P_{N-1]}(X^N \cdot \Phi_0)\| \neq 0 \end{cases}$$

(6)

and the orthogonal projection

$$P_{N]}(\xi) := \sum_{j=0}^{N} \langle \Phi_j, \xi \rangle \Phi_j.$$

(7)

Clearly either $\|\Phi_j\| = 1$ or $\|\Phi_j\| = 0$ and, since by the induction assumption (5) the Φ_j ($j \in \{0, \ldots, N-1\}$) are real polynomials and X is symmetric, one has

$$\langle X^N \cdot \Phi_0, \Phi_j \rangle = \langle \Phi_0, X^N \cdot \Phi_j \rangle = \int_{\mathbb{R}} x^N \Phi_j(x) \mu(dx) \in \mathbb{R}.$$

This implies that the scalar products

$$\langle \Phi_j, X^N \cdot \Phi_0 \rangle \qquad ; \qquad j \in \{0, 1, \ldots, N-1\}$$

are real numbers and that the polynomial

$$P_{N-1]}(X^N \cdot \Phi_0) = \sum_{j=0}^{N} \langle \Phi_j, X^N \cdot \Phi_0 \rangle \Phi_j$$

has real coefficients. Then (6) implies that the same is true for Φ_N. More explicitly

$$\Phi_N = c_N (X^N + Q_{N-1}) \cdot \Phi_0$$

where

$$c_N = \begin{cases} 1 \ , \ \text{if} \ \|\Phi_N\| = 0 \\ \|X^N \cdot \Phi_0 - P_{N-1]}(X^N \cdot \Phi_0)\|^{-1} \ \text{if} \ \|\Phi_N\| \neq 0. \end{cases}$$

(8)

Since, by construction the linear span of $(\Phi_j)_{j=1}^{N}$ coincides with the linear span of the set

$$\{X^N \cdot \Phi_0\} \cup \mathcal{P}_{N-1]} \cdot \Phi_0 = \mathcal{P}_{N]} \cdot \Phi_0.$$

Notice that, if $\|\Phi_j\| = 0$ then, for any vector $\xi \in \mathcal{P} \cdot \Phi_0$

$$|\langle \xi, \Phi_j \rangle| \leq \|\xi\| \cdot \|\Phi_j\| = 0$$

i.e. Φ_j is orthogonal to all vectors in $\mathcal{P} \cdot \Phi_0$. Therefore each $P_{n]}$ ($n \in \{0, \ldots, N\}$) is effectively an orthogonal projection. This defines by induction the sequences $(\Phi_n)_{n \in \mathbb{N}}$ and $(P_{n]})$.

3.1. *The μ-orthogonal gradation of $\mathcal{P}$*

Since, for each $n \in \mathbb{N}$, the range of $P_{n]}$ is $\mathcal{P}_{n]}$, it follows that

$$P_{n]} \leq P_{n+1]} \quad ; \quad \lim_{n \to \infty} P_{n]} = 1_{\mathcal{P}}$$

where the limit is meant in the sense that

$$\forall Q \in \mathcal{P}, \exists n_Q \in \mathbb{N} : P_{n]}(Q) = P_{n_Q]}(Q) \quad ; \quad \forall n \geq n_Q.$$

Define

$$P_n := P_{n]} - P_{n-1]}.$$

Then for any $m, n \in \mathbb{N}$

$$m \neq n \Rightarrow P_m P_n = 0$$

$$P_n = P_n^*$$

$$\sum_{n \in \mathbb{N}} P_n = 1_{\mathcal{P}}.$$

Define

$$\mathcal{P}_n := P_n(\mathcal{P}) = \text{Range } (P_n) = \mathbb{C} \cdot \Phi_n.$$

Then

$$\mathcal{P} = \bigoplus_{n \in \mathbb{N}} \mathcal{P}_n = \bigoplus_{n \in \mathbb{N}} \mathbb{C} \cdot \Phi_n \tag{9}$$

where the symbol $\bigoplus$ denotes orthogonal sum in the pre-Hilbert space $\mathcal{P}$. This decomposition is called **the μ-orthogonal gradation** of $\mathcal{P}$.

Theorem 3.1: *Suppose that there exists $N \in \mathbb{N}$ such that*

$$\|X^N \cdot \Phi_0 - P_{N-1]}(X^N \cdot \Phi_0)\| = 0.$$

Then

$$\|X^{N+k} \cdot \Phi_0 - P_{N-1]}(X^{N+k} \cdot \Phi_0)\| = 0 \quad ; \quad \forall k \in \mathbb{N}. \tag{10}$$

Proof: Denote

$$\mathcal{N}_\mu := \{\xi \in \mathcal{P} \cdot \Phi : \|\xi\| = 0\} \tag{11}$$

we say that two vectors $\xi, \eta \in \mathcal{P}$ are equivalent and we write $\xi \equiv \eta$, if

$$\|\xi - \eta\| = 0 \quad (\Leftrightarrow \xi \in \eta + \mathcal{N}_\mu).$$

The fact that

$$\|X^N \cdot \Phi_0 - P_{N-1]}(X^N \cdot \Phi_0)\| = 0$$

means that

$$X^N \cdot \Phi_0 \in P_{N-1]}(X^N \cdot \Phi_0) + \mathcal{N}_\mu \subseteq \mathcal{P}_{N-1]} \cdot \Phi_0 + \mathcal{N}_\mu$$

and this implies that

$$\mathcal{P}_{N]} \cdot \Phi_0 = \mathbb{C} \cdot X^N \cdot \Phi_0 \oplus \mathcal{P}_{N-1]} \cdot \Phi_0 \subseteq \mathcal{P}_{N-1]} \cdot \Phi_0 + \mathcal{N}_\mu.$$

Since, by the Schwarz inequality, $X\mathcal{N}_\mu \subseteq \mathcal{N}_\mu$, this implies that

$$X\mathcal{P}_{N]} \cdot \Phi_0 \subseteq X\mathcal{P}_{N-1]} \cdot \Phi_0 + \mathcal{N}_\mu \subseteq \mathcal{P}_{N]} \cdot \Phi_0 + \mathcal{N}_\mu \subseteq \mathcal{P}_{N-1]} \cdot \Phi_0 + \mathcal{N}_\mu.$$

Therefore

$$\mathcal{P}_{N+1]} \cdot \Phi_0 = X\mathcal{P}_{N]} \cdot \Phi_0 + \mathcal{P}_{N]} \cdot \Phi_0 \subseteq \mathcal{P}_{N-1]} \cdot \Phi_0 + \mathcal{N}_\mu$$

and, by induction,

$$\mathcal{P}_{N+k]} \cdot \Phi_0 \subseteq \mathcal{P}_{N-1]} \cdot \Phi_0 + \mathcal{N}_\mu \ ; \ \forall k \in \mathbb{N}.$$

This implies that, for some $n_\mu \in \mathcal{N}_\mu$

$$X_{N+k]} \cdot \Phi_0 = Q_{N-1} \cdot \Phi_0 + n_\mu$$

i.e. that

$$\|X_{N+k]} \cdot \Phi_0 - Q_{N-1} \cdot \Phi_0\| = 0.$$

Therefore

$$0 = \|(X^{N+k} \cdot \Phi_0 - P_{N-1]}(X^{N+k} \cdot \Phi_0)) + (P_{N-1]}(X^{N+k} \cdot \Phi_0) - Q_{N-1} \cdot \Phi_0)\|$$

$$= \|(X^{N+k} \cdot \Phi_0 - P_{N-1]}(X^{N+k} \cdot \Phi_0)) + (P_{N-1]}(X^{N+k} \cdot \Phi_0 - Q_{N-1} \cdot \Phi_0))\| = 0.$$

Since

$$\|P_{N-1]}(X^{N+k} \cdot \Phi_0 - Q_{N-1} \cdot \Phi_0))\| \le \|X^{N+k} \cdot \Phi_0 - Q_{N-1} \cdot \Phi_0)\| = 0$$

it follows that

$$0 = \|(X^{N+k} \cdot \Phi_0 - P_{N-1]}(X^{N+k} \cdot \Phi_0)) + (P_{N-1]}(X^{N+k} \cdot \Phi_0 - Q_{N-1} \cdot \Phi_0))\|^2$$

$$= \|X^{N+k} \cdot \Phi_0 - P_{N-1]}(X^{N+k} \cdot \Phi_0)\|^2 + \|P_{N-1]}(X^{N+k} \cdot \Phi_0 - Q_{N-1} \cdot \Phi_0)\|^2$$

which implies that

$$X^{N+k} \cdot \Phi_0 = P_{N-1]}(X^{N+k} \cdot \Phi_0) + \mathcal{N}_\mu.$$

Since $k \ge 0$ is arbitrary, this is equivalent to (10).

Corollary 3.1: *If there exists $N \in \mathbb{N}$ such that $\Phi_N = 0$, then*

$$\Phi_{N+k} = 0 \quad ; \quad \forall k \in \mathbb{N}.$$

Proof: This is clear from the definition of the Φ_j.

Theorem 3.2: *For the μ-orthogonal gradation of $\mathcal{P}$ the following alternative holds:*
(i) Either

$$\mathcal{P} = \bigoplus_{n \in \mathbb{N}} \mathbb{C} \cdot \Phi_n$$

with

$$\|\Phi_n\| = 1 \quad ; \quad \forall n \in \mathbb{N}$$

and in this case μ has infinite support.

(ii) Or there exists $n_\mu \in \mathbb{N}$ such that

$$\mathcal{P} = \bigoplus_{n \in \{1,\dots,n_\mu\}} \mathbb{C} \cdot \Phi_n \oplus \bigoplus_{n > n_\mu} \mathbb{C} \cdot \Phi_n$$

with

$$\|\Phi_n\| = 0 \quad ; \quad \forall n > n_\mu$$

and in this case μ is a finite convex combination of Dirac measures.

Definition 3.1: In the notation (11), we say that two linear operators

$$A, B : \mathcal{P} \to \mathcal{P}$$

are μ-equivalent and we write $A \equiv_\mu B$, if

$$\|A\xi - B\xi\| = 0 \quad ; \quad \forall \xi \in \mathcal{P}.$$

3.2. The symmetric Jacobi relation

$$\sum_{n \in \mathbb{N}} P_n = 1_\mathcal{P}$$

$$X = 1_\mathcal{P} X 1_\mathcal{P} = \sum_{m,n \in \mathbb{N}} P_m X P_n.$$

Therefore

$$X P_n = \sum_{m \in \mathbb{N}} P_m X P_n.$$

Theorem 3.3: *(Jacobi 1870). For any* $n \in \mathbb{N}$

$$XP_n = P_{n+1}XP_n + P_nXP_n + P_{n-1}XP_n.$$

*This identity is called **the symmetric Jacobi relation.***

Proof: Let $P_n \in \mathcal{P}_n$ and $P_{n+k} \in \mathcal{P}_{n+k}$. Then, if $k > 1$

$$\langle X\Phi_n, \Phi_{n+k} \rangle = 0$$

because $\Phi_{n+k} \perp \mathcal{P}_{n+k-1]} \supseteq \mathcal{P}_{n+1]}$. This is equivalent to:

$$k > 1 \Rightarrow P_{n+k}XP_n = 0.$$

Taking the adjoint of this one finds

$$k > 1 \Rightarrow P_nXP_{n+k} = 0$$

and this proves the statement.

3.3. *Creation and annihilation operators*
Define $\forall n \in \mathbb{N}$

$$a_n^+ := P_{n+1}XP_n|_{\mathcal{P}_n} : \mathbb{C} \cdot \Phi_n \to \mathbb{C} \cdot \Phi_{n+1}$$

$$a_n^0 := P_nXP_n|_{\mathcal{P}_n} : \mathbb{C} \cdot \Phi_n \to \mathbb{C} \cdot \Phi_n$$

$$a_n^- := P_{n-1}XP_n|_{\mathcal{P}_n} : \mathbb{C} \cdot \Phi_n \to \mathbb{C} \cdot \Phi_{n+1}.$$

Then the symmetric Jacobi relation becomes

$$XP_n = a_n^+ + a_n^0 + a_n^-$$

and implies

$$(a_n^0)^* = a_n^0 \quad ; \quad (a_n^+)^* = a_n^-.$$

Summing over n one finds

$$X = \sum_{n \in \mathbb{N}} a_n^+ + \sum a_n^0 + \sum a_n^- =: A^+ + A^0 + A^-$$

$$(A^0)^* = A^0 \quad ; \quad (A^+)^* = A^-.$$

This is called **the quantum decomposition of** X.

4. The 1-Dimensional Case

For $\Phi \in \mathcal{P} \cdot \Phi_0$, we use the notation ('bra' vector):

$$\Phi^* : \mathcal{P} \cdot \Phi_0 \to \mathbb{C}$$

$$\Phi^*(\xi) := \langle \Phi, \xi \rangle.$$

Recall that

$$P_n := P_{n]} - P_{n-1]} = \langle \Phi_n, \cdot \rangle \Phi_n.$$

Therefore in the 1-dimensional case the symmetric Jacobi relation is equivalent to

$$X \Phi_m \Phi_n^* = \langle \Phi_{n+1}, X \Phi_n \rangle \Phi_{n+1} \Phi_n^* + \langle \Phi_n, X \Phi_n \rangle \Phi_n \Phi_n^* + \langle \Phi_{n-1}, X \Phi_n \rangle \Phi_{n-1} \Phi_n^*. \tag{12}$$

Define

$$\beta_n := \langle \Phi_{n-1}, X \Phi_n \rangle \in \mathbb{R}$$

$$\alpha_n := \langle \Phi_n, X \Phi_n \rangle \in \mathbb{R}$$

$\beta_n \in \mathbb{R}$ because we have proved that the Φ_n are polynomials with real coefficients and we know that the scalar product of 2 such polynomials is a real number.

Lemma 4.1: *If $\beta_n = 0$, then*

$$\beta_k = 0 \quad ; \quad \forall k \geq n.$$

Proof: Suppose that $\beta_n = 0$, then

$$\beta_{n+1} = \langle \Phi_n, X \Phi_{n+1} \rangle$$

$$= \langle X \Phi_n, \Phi_{n+1} \rangle = \langle \alpha_n \Phi_n + \beta_{n-1} \Phi_{n-1}, \Phi_{n+1} \rangle = 0.$$

Because Φ_{n+1} is orthogonal to $\mathcal{P}_{n]}$.

This implies in particular that

$$\langle \Phi_{n-1}, X \Phi_n \rangle = \langle X \Phi_n, \Phi_{n-1} \rangle = \langle \Phi_n, X \Phi_{n-1} \rangle = \beta_{n-1}.$$

Where, by definition

$$\beta_{-1} := 0.$$

With these notations, if $||\Phi_n|| = 0$ then the right-hand side of (12) is identically zero. If $||\Phi_n|| = 1$ then, evaluating both sides in Φ_n, one finds

$$X\Phi_n = \beta_n \Phi_{n+1} + \alpha_n \Phi_n + \beta_{n-1}\Phi_{n-1}.$$

This is the symmetric 1-dimensional Jacobi relation. With these notations summing over $n \in \mathbb{N}$ one obtains

$$X = A^+ + A^0 + A^-$$

$$A^+ = \sum_{n \in \mathbb{N}} \beta_n \Phi_{n+1}\phi_n^*$$

$$A^0 = \sum \alpha_n \Phi_n \Phi_n^*$$

$$A^- = \sum \beta_{n-1}\Phi_{n-1}\Phi_n^*.$$

More explicitly

$$A^+\Phi_n = \beta_n \Phi_{n+1}$$

$$A^0\Phi_n = \alpha_n \Phi_n$$

$$A^-\Phi_n = \beta_{n-1}\Phi_{n-1}.$$

Therefore $\forall n \in \mathbb{N}$

$$A^- A^+ \Phi_n = \beta_n A^- \Phi_{n+1} = \beta_n^2 \Phi_n$$

$$A^+ A^- \Phi_n = \bar{\beta}_{n-1} A^+ \Phi_{n-1} = \beta_{n-1}^2 \Phi_n$$

$$[A^-, A^+]\Phi_n = (\beta_n^2 - \beta_{n-1}^2)\Phi_n.$$

Thus defining **the principal Jacobi sequence**

$$\omega_n := \beta_n^2 \quad ; \quad n \in \mathbb{N} \tag{13}$$

and the **number operator**

$$\Lambda\Phi_n := n\Phi_n \; ; \; n \in \mathbb{N} \tag{14}$$

$$\Lambda = \sum_{n \in \mathbb{N}} n\Phi_n \Phi_n^*$$

we obtain

$$[A^-, A^+] = \omega_\Lambda - \omega_{\Lambda-1} = \sum_{n \in \mathbb{N}} (\omega_n - \omega_{n-1})\Phi_n \Phi_n^*.$$

4.1. *Monic Jacobi relations*

From now on we assume that

$$\Phi_n \neq 0 \quad ; \quad \forall n \in \mathbb{N}.$$

This is equivalent to say that the support of the measure(s) associated to the state φ is not reduced to a finite number of points or that

$$\omega_n > 0 \quad ; \quad \forall n \in \mathbb{N}.$$

In this case

$$\Phi_n := \frac{\Phi_n^0}{\|\Phi_n^0\|} \quad ; \quad \forall n$$

and one can always assume that the polynomial Φ_n^0 is monic. Dividing both sides of the identity

$$X\Phi_n = \beta_n \Phi_{n+1} + \alpha_n \Phi_n + \beta_{n-1}\Phi_{n-1}$$

by $\|\bar{\Phi}_n^0\|$ one finds:

$$X\Phi_n^0 = \beta_n \frac{\|\Phi_n^0\|}{\|\Phi_{n+1}^0\|}\Phi_{n+1}^0 + \alpha_n \Phi_n^0 + \beta_{n-1}\frac{\|\Phi_n^0\|}{\|\Phi_{n-1}^0\|}.$$

The fact that the polynomial Φ_n^0 is monic implies that

$$\beta_n \frac{\|\Phi_n^0\|}{\|\Phi_{n+1}^0\|} = 1 \Leftrightarrow \beta_n = \frac{\|\Phi_{n+1}^0\|}{\|\Phi_n^0\|}.$$

In particular (additional proof that the β_n are real and in fact positive)

$$\beta_n \geq 0.$$

Therefore

$$\frac{\|\Phi_n^0\|}{\|\Phi_{n-1}^0\|} = \beta_{n-1}$$

and the Symmetric Jacobi relations become equivalent to

$$\Leftrightarrow X\Phi_n^0 = \beta_n \frac{\|\Phi_n^0\|}{\|\Phi_{n+1}^0\|}\Phi_{n+1}^0 + \alpha_n \Phi_n^0 + \beta_{n-1}\frac{\|\Phi_n^0\|}{\|\Phi_{n-1}^0\|}$$

$$\Leftrightarrow X\Phi_n^0 = \Phi_{n+1}^0 + \alpha_n \Phi_n^0 + \beta_{n-1}^2 \Phi_{n-1}^0.$$

4.2. *Commutation relations in monic form*

Recalling the definition (13) of the ω_n, the symmetric Jacobi relations

$$X\Phi_n = \beta_n \Phi_{n+1} + \alpha_n \Phi_n + \beta_{n-1}\Phi_{n-1}$$

become **the monic Jacobi relations**

$$X\Phi_n^0 = \Phi_{n+1}^0 + \alpha_n \Phi_n^0 + \omega_n \Phi_{n-1}^0.$$

This is the form of the Jacobi relations found in the books. From Lemma 4.1 we know that

$$\omega_n = 0 \ \Rightarrow\ \omega_{n+k} = 0 \ \ ; \ \ \forall k \geq 0.$$

The sequence (ω_n) is called **the principal Jacobi sequence.**

The sequence (α_n) is called **the secondary Jacobi sequence.** The monic form of the commutation relations is then:

$$[a^-, a^+] = (\beta_\Lambda^2 - \beta_{\Lambda-1}^2) = (\omega_\Lambda - \omega_{\Lambda-1}) = \sum(\omega_n - \omega_{n-1})\Phi_n\Phi_n^*.$$

The commutation relation

$$[a^-, a^+] = (\omega_\Lambda - \omega_{\Lambda-1})$$

are called **the probabilistic commutation relations.**

4.3. *The Gaussian case*

It is known that the principal Jacobi sequence for the standard Gaussian measure on $\mathbb{R}$ are given by

$$\omega_n = n \ \ ; \ \ \alpha_n = 0.$$

With this choice the commutation relations become

$$[a^-, a^+] = (\omega_\Lambda - \omega_{\Lambda-1}) = (\Lambda - (\Lambda - 1)) = id_\mathcal{P} =: 1.$$

These are the **Heisenberg commutation relations**

$$[a^-, a^+] = 1.$$

If, instead of the Gaussian, one chooses other measures one obtains different generalizations of the Heisenberg commutation relations.

The following table provides some examples.

Measure	Polynomials	Jacobi parameters
Gaussian $N(0,\sigma^2)$	Hermite $H_n(x;\sigma^2)$ $= (-\sigma^2)^n e^{x^2/2\sigma^2}\partial_x^n e^{-x^2/2\sigma^2}$	$\alpha_n = 0, \ \omega_n = \sigma^2 n$
Poisson Poi(a) $(-1)^n a^{-x}\Gamma(x+1)\Delta^n\left[\frac{a^x}{\Gamma(x-n+1)}\right]$	Charlier $C_n(x;a) =$	$(\lambda_n = \sigma^{2n}n!)$
Gamma $\Gamma(\alpha)$, $(\alpha > -1)$ $\frac{1}{\Gamma(\alpha+1)}x^\alpha e^{-x}, x>0$	Laguerre $\mathcal{L}_n^{(\alpha)}(x)$ $= (-1)^n x^{-\alpha}e^x \partial_x^n[x^{n+\alpha}e^{-x}]$	$\alpha_n = n+a, \ \omega_n = an$ $(\lambda_n = a^n n!)$ $\alpha_n = 2n+1+\alpha, \ \omega_n = n(n+\alpha)$ $(\lambda_n = n!(n+\alpha)\cdots(1+\alpha))$
Uniform on $[-1,1]$	Legendre $\tilde{L}_n(x) = \frac{1}{2^n(2n-1)!!}\partial_x^n[(x^2-1)^n]$	$\alpha_n = 0, \ \omega_n = \frac{n^2}{(2n+1)(2n-1)}$ $\left(\lambda_n = \frac{(n!)^2}{[(2n-1)!!]^2(2n+1)}\right)$
Arcsine $\frac{1}{\pi\sqrt{1-x^2}}, \ |x|<1$	Chebyshev (1st kind) $\tilde{T}_0(x) = 1$ $\tilde{T}_n(x) = \frac{1}{2^{n-1}}\cos(n\cos^{-1}x), n\geq 1$	$\alpha_n = 0 \ \ \omega_n = \begin{cases}\frac12 & n=1 \\ \frac14 & n\geq 2\end{cases}$ $\left(\lambda_n = \frac{1}{2^{2n-1}}\right)$
Semicircle $\frac{2}{\pi}\sqrt{1-x^2}, |x|<1$	Chebyshev (2nd kind) $\tilde{U}_n(x) = \frac{1}{2^n}\frac{\sin[(n+1)\cos^{-1}x]}{\sin[\cos^{-1}x]}$	$\alpha_n = 0, \ \omega_n = 1/4$ $\left(\lambda_n = \frac{1}{4^n}\right)$
Beta $\frac{1}{\sqrt{\pi}}\frac{\Gamma(\beta+1)}{\Gamma\left(\beta+\frac12\right)}(1-x^2)^{\beta-\frac12}$ $|x|<1, \beta > -\frac12$	Gegenbauer $\tilde{G}_n^{(\beta)}(x) = C_n^{(\beta)}(1-x^2)^{\frac12-\beta}\partial_x^n[(1-x^2)^{n+\beta-\frac12}]$ $C_n^{(\beta)} = \frac{(-1)^n 2^n\Gamma(2\beta+n)}{\Gamma(2\beta+2n)}$	$\alpha_n = 0, \ \omega_n = \frac{n(n+2\beta-1)}{4(n+\beta)(n+\beta-1)}$

$[a^-,a^+]e_n$	Coherent vector	Generating function
$\sigma^2 I$	$e^{\frac{zx}{\sigma^2}-\frac{z^2}{2\sigma^2}}$	$e^{tx-\frac12\sigma^2 t^2} = \sum_{n=0}^\infty \frac{H_n(x;\sigma^2)}{n!}t^n$
aI	$e^{-z}\left(1+\frac{z}{a}\right)^x$	$e^{-at}(1+t)^x = \sum_{n=0}^\infty \frac{C_n(x;a)}{n!}t^n$
$(1+t)^{-\alpha-1}e^{\frac{tx}{1+t}}(2n+\alpha+1)e_n$	$\sum_{n=0}^\infty \frac{\mathcal{L}_n^{(\alpha)}(x)}{n!(n+\alpha)\cdots(1+\alpha)}z^n$	$= \sum_{n=0}^\infty \frac{t^n}{n!}\mathcal{L}_n^{(\alpha)}(x)$
$\frac{1}{\sqrt{1-2tx+t^2}} - \frac{1}{(2n+3)(2n+1)(2n-1)}e_n$	$\sum_{n=0}^\infty \frac{((2n-1)!!)^2(2n+1)}{(n!)^2}\tilde{L}_n(x)z^n$ $= \sum_{n=0}^\infty \frac{(2n-1)!!}{n!}\tilde{L}_n(x)t^n$	
$\begin{cases}\frac12 e_0, & n=0 \\ -\frac14 e_1, & n=1 \\ 0, & n\geq 2\end{cases}$	$\frac{1-2xz}{1-4xz+4z^2}$ $= \sum_{n=0}^\infty \tilde{T}_n(x)t^n$	$\frac{4-t^2}{4-4tx+t^2}$
$\begin{cases}\frac14 e_0, & n=0 \\ 0, & n\geq 1\end{cases}$	$\frac{1}{1-4xz+4z^2}$	$\frac{4}{4-4tx+t^2} = \sum_{n=0}^\infty \tilde{U}_n(x)t^n$
$\frac{1}{(1-2tx+t^2)^\beta}$	$\frac{\beta^2-\beta}{2(n+1+\beta)(n+\beta)(n-1+\beta)}$	not in closed form $= \sum_{n=0}^\infty \frac{2^n\Gamma(\beta+n)}{\Gamma(\beta)n!}\tilde{G}_n^{(\beta)}(x)t^n$

5. Probabilistic Extensions of Quantum Mechanics

5.1. *The generalized free evolution*

We consider the symmetric case:

$$a^0 = 0.$$

Start from the probabilistic commutation relations in monic form:

$$[a^-,a^+] = \omega_\Lambda - \omega_{\Lambda-1}$$

$$[a^+,a^+] = [a^-,a^-] = 0.$$

The Schrödinger equation (with $\hbar = 1$) is:

$$\partial_t\psi_t = -iH\psi_t. \tag{15}$$

Taking

$$H := ca^+a^- \quad ; \quad c > 0$$

the corresponding Heisenberg evolution for $a^\pm$ is

$$a_t := u_t(a^\pm) := e^{itH}a^\pm e^{-itH} = e^{itca^+a^-}a^\pm e^{-itca^+a^-}. \tag{16}$$

In the notations (13), (14), from

$$[a^-, a^+a^-] = [a^-, a^+]a^- = (\omega_\Lambda - \omega_{\Lambda-1})a^-$$

and the fact that, since ca^+a^- leaves the orthogonal gradation invariant, it commutes with Λ,

$$[a^+a^-, \Lambda] = 0$$

and, since a^+a^- and Λ have discrete spectrum, with all its Borel functions, one deduces that

$$\frac{d}{dt}a_t = \frac{d}{dt}e^{itca^+a^-}a^- e^{-itca^+a^-} = ite^{itca^+a^-}[ca^+a^-, a^-]e^{-itca^+a^-}$$

$$= -ite^{itca^+a^-}c(\omega_\Lambda - \omega_{\Lambda-1})a^- e^{-itca^+a^-}$$

$$= -itc(\omega_\Lambda - \omega_{\Lambda-1})e^{itca^+a^-}a^- e^{-itca^+a^-}.$$

Therefore, denoting

$$a^-(t) := e^{itca^+a^-}a^- e^{-itca^+a^-}$$

since $a^-(0) = a^-$ one obtains

$$\frac{d}{dt}a^-(t) = -it(\omega_\Lambda - \omega_{\Lambda-1})a^-(t)$$

$$a^-(0) = a^-$$

whose unique solution is

$$a^-(t) = e^{itca^+a^-}a^- e^{-itca^+a^-} = e^{-itc(\omega_\Lambda - \omega_{\Lambda-1})}a^-.$$

Therefore

$$a^+(t) = a^+ e^{itc(\omega_\Lambda - \omega_{\Lambda-1})}.$$

This implies, using again

$$[a^+a^-, \Lambda] = [a^-a^+, \Lambda] = 0$$

that

$$[a(t), a^+(t)]$$

$$= e^{-itc(\omega_\Lambda - \omega_{\Lambda-1})}a^- a^+ e^{itc(\omega_\Lambda - \omega_{\Lambda-1})} - a^+ e^{itc(\omega_\Lambda - \omega_{\Lambda-1})}e^{-itc(\omega_\Lambda - \omega_{\Lambda-1})}a^-$$

$$= a^- a^+ - a^+ a^- = [a, a^+].$$

Thus the map

$$t \in R \;\mapsto\; a^\pm(t) = \begin{cases} a^+ e^{it(\omega_\Lambda - \omega_{\Lambda-1})} \\ e^{-it(\omega_\Lambda - \omega_{\Lambda-1})}a^- \end{cases} \tag{17}$$

called **the generalized free evolution**, is a $*$-Lie-algebra isomorphism, hence it extends to an (associative) $*$-algebra isomorphism of the universal enveloping algebra of $(a^+, a^-, 1)$. This means that the $*$-algebra

$$\mathrm{Pol}(a^\pm) := \text{algebraic span of}\{a^\pm\}$$

is left invariant by its unique $*$-automorphism extending the generalized free evolution.

5.2. *Equilibrium states*

From the expression of the generalized free evolution

$$u_t(X) = e^{it(\omega_\Lambda - \omega_{\Lambda-1})}X e^{-it(\omega_\Lambda - \omega_{\Lambda-1})} \qquad ; \qquad t \in R$$

by analytic continuation at $t = -i\beta$ one obtains

$$u_{-i\beta}(X) = e^{\beta(\omega_\Lambda - \omega_{\Lambda-1})}X e^{-\beta(\omega_\Lambda - \omega_{\Lambda-1})}. \tag{18}$$

The KMS equilibrium condition at inverse temperature β, for a given density matrix W, is:

$$\mathrm{Tr}(W u_{t-i\beta}(X)Y) = \mathrm{Tr}(W Y u_t(X)).$$

Putting $t = 0$ this becomes

$$\mathrm{Tr}(W u_{-i\beta}(X)Y) = \mathrm{Tr}(W Y X).$$

Using the explicit form (18) of $u_{-i\beta}(X)$, this is equivalent to

$$\mathrm{Tr}(W Y X)\mathrm{Tr}(X W Y) = \mathrm{Tr}(W u_{-i\beta}(X)Y)$$

$$= \mathrm{Tr}(W e^{\beta(\omega_\Lambda - \omega_{\Lambda-1})}X e^{-\beta(\omega_\Lambda - \omega_{\Lambda-1})}Y).$$

The arbitrariness of Y implies that

$$\Leftrightarrow W e^{\beta(\omega_\Lambda - \omega_{\Lambda-1})} X e^{-\beta(\omega_\Lambda - \omega_{\Lambda-1})} = XW$$

$$\Leftrightarrow W e^{\beta(\omega_\Lambda - \omega_{\Lambda-1})} X = XW e^{\beta(\omega_\Lambda - \omega_{\Lambda-1})}.$$

Since X is arbitrary, $W e^{\beta(\omega_\Lambda - \omega_{\Lambda-1})}$ must be a multiple of the identity:

$$W e^{\beta(\omega_\Lambda - \omega_{\Lambda-1})} =: \frac{1}{Z_\beta} \cdot 1 \Leftrightarrow W = \frac{e^{-\beta(\omega_\Lambda - \omega_{\Lambda-1})}}{Z_\beta}.$$

References

1. L. Accardi and M. Bożejko,
   Interacting Fock space and Gaussianization of probability measures,
   IDA–QP (Infin. Dim. Anal. Quantum Probab. Rel. Topics) 1 (1998) 663-670.
2. Accardi L., Barhoumi A., Dhahri A.,
   Identification of the theory of multi–dimensional orthogonal polynomials with
   the theory of symmetric interacting Fock spaces with finite dimensional 1–
   particle space.
   submitted for publication November 2013.
   https://www.scienceopen.com/document/vid/fb3fd593-cb8d-43a2-837d-
   eb2d2508669f
3. L. Accardi, H.-H. Kuo and A.I. Stan,
   Characterization of probability measures through the canonically associated
   interacting Fock spaces,
   IDA–QP (Inf. Dim. Anal. Quant. Prob. Rel. Top.) 7 (4) (2004), 485-505.
4. L. Accardi, H.-H. Kuo and A. Stan,
   Probability measures in terms of creation, annihilation, and neutral operators,
   Quantum Probability and Infinite Dimensional Analysis: From Foundations
   to Applications. [QP-PQ XVIII], M. Schürmann and U. Franz (eds.), World
   Scientific (2005), 1-11.
5. L. Accardi, H.-H. Kuo and A. Stan,
   Moments and commutators of probability measures,
   IDA–QP (Infin. Dim. Anal. Quantum Probab. Rel. Topics) 10 (4) (2007),
   591-612.
6. Accardi L., Nhani M.,
   The interacting Fock space of Haldane's exclusion statistics,
   in: Proceedings of the 2–d international symposium on Quantum Theory and
   Symmetries, Krakow, 18–21 (2001), E. Kapuscik, A. Horzela (eds.), World
   Scientific (2001) 1–10.
   Preprint Volterra, N. 491 (2001).

7. Accardi L., Nhani M.,
   Interacting Fock Spaces and Orthogonal Polynomials in several variables,
   in: "Non-Commutativity, Infinite-Dimensionality and Probability at the
   Crossroad" N. Obata, A. Hora, T. Matsui (eds.), World Scientific (2002)
   192–205.
   Preprint Volterra, N. 523 (2002).

HIDA DISTRIBUTION CONSTRUCTION OF INDEFINITE METRIC $(\phi^p)_d$ $(d \geq 4)$ QUANTUM FIELD THEORY

Sergio Albeverio

Angewandte Mathematik and HCM, Universität Bonn
Endericherallee 60, D-53115 Bonn, Germany;
SFB611; IZKS; BiBoS; CERFIM, Locarno;
Ist. Matematica, Università di Trento, Italy
albeverio@uni-bonn.de

Minoru W. Yoshida

Department of Mathematics, Tokyo City University
1-28-1 Tamazutsumi Setagaya Tokyo 158-8557, Japan
myoshi@tcu.ac.jp

For the space-time dimension $d = 4$, analogous to the case of $d = 2$, we consider the Feynman graphs that may appear $(\phi^p)_4$ Euclidean quantum field theory without cutoff. We see that some of the graphs, which are not the ones of the free field, are able to be defined as *Hida distributions*. The present discussion is a generalization of [AY3].

1. Short Review of a Probabilistic Formulation of Euclidean $(\phi^4)_2$

Throughout this chapter we let $d = 4$ or $d = 2$ with an adequate understanding. Let $\dot{W}$ be the random variable such that $\dot{W}(\omega) \in \mathcal{S}'(\mathbb{R}^d \to \mathbb{R})$, $P - a.e.$ $\omega \in \Omega$, and for each $\varphi \in \mathcal{S}(\mathbb{R}^d \to \mathbb{R})$, $< \dot{W}, \varphi >_{\mathcal{S}',\mathcal{S}}$ is a real valued Gaussian random variable (*white noise on $\mathbb{R}^d$*) satisfying for any φ_1, $\varphi_2 \in \mathcal{S}(\mathbb{R}^d \to \mathbb{R})$

$$E\left[< \dot{W}, \varphi >_{\mathcal{S}',\mathcal{S}}\right] = 0, \tag{1.1}$$

$$E\left[< \dot{W}, \varphi_1 >_{\mathcal{S}',\mathcal{S}} \cdot < \dot{W}, \varphi_2 >_{\mathcal{S}',\mathcal{S}}\right] = \int_{\mathbb{R}^d} \varphi_1(\mathbf{x})\, \varphi_2(\mathbf{x})\, d\mathbf{x}. \tag{1.2}$$

Let $J_{d=2}^{\frac{1}{2}}$ be the integral kernel of the pseudo differential operator on $\mathcal{S}(\mathbb{R}^2)$ such that $(-\Delta + 1)^{-\frac{1}{2}}$ with $\Delta = \Delta_{d=2}$ the Laplace operator on $\mathbb{R}^2$. For φ, $f_j \in \mathcal{S}(\mathbb{R}^2 \to \mathbb{R})$, $j = 1, \ldots, n$, let

$$\phi(f_j) = \int_{\mathbb{R}^2} \left(\int_{\mathbb{R}^2} f_j(\mathbf{x}) \, J_{d=2}^{\frac{1}{2}}(\mathbf{x} - \mathbf{y}) d\mathbf{x} \right) \dot{W}(\mathbf{y}) d\mathbf{y}, \qquad (1.3)$$

which is known as the 2-dimensional Euclidean *free* field operator. For $p = 2q$ (i.e., an even number), $q \in \mathbb{N} =$ the set of natural numbers, and bounded $\Lambda \subset \mathbb{R}^2$, let

$$<: \phi_{d=2}^{2q} :, \varphi > \qquad (1.4)$$

$$= \int_{(\mathbb{R}^2)^{2q}} \left\{ \int_{\mathbb{R}^2} \varphi(x) \prod_{i=1}^{2q} J_{d=2}^{\frac{1}{2}}(x - y_i) dx \right\} : \dot{W}(\mathbf{y}_1) \cdots \dot{W}(\mathbf{y}_{2q}) : \quad (1.5)$$

$$\times d\mathbf{y}_1 \cdots d\mathbf{y}_{2q}. \qquad (1.6)$$

Then,

$$e^{-\lambda <: \phi_{d=2}^{2q} :, 1_\Lambda >} \in \cap_{p \geq 1} L^p(\Omega; P), \qquad (1.7)$$

and the Schwinger function for $d = 2$

$$S_n(f_1, \ldots, f_n) \equiv \frac{1}{Z(\lambda; \Lambda)} E\left[\phi(f_1) \cdots \phi(f_n) e^{-\lambda <: \phi_{d=2}^{2q} :, 1_\Lambda >} \right], \qquad n \in \mathbb{N}, \qquad (1.8)$$

is well defined as the expectations. S_n is known as the Schwinger function of 2-dimensional $(\phi^{2q})_2$ quantum field theory with the cutoff $\Lambda \subset \mathbb{R}^2$.

2. Formulation for $d = 4$

By changing the 2-dimensional space-time Gaussian white noise process by 4-dimensional space-time ones in the previous discussion, and if we apply the same considerations to $(\Phi^{2q})_4$ quantum field model directly, then the terms $\phi(f_1) \cdots \phi(f_n) (<: \phi_{d=4}^{2q} :, 1_\Lambda >)^k$ will not have the right to be random variables. We need an approximation method.

Denote

$$\mathbf{x} \equiv (t, \overrightarrow{x}) \in \mathbb{R} \times \mathbb{R}^3 = \mathbb{R}^4, \qquad \xi \equiv (\tau, \overrightarrow{\xi}) \in \mathbb{R} \times \mathbb{R}^3 = \mathbb{R}^4.$$

Let

$$\mathcal{F}[\varphi](\xi) = \int_{\mathbb{R}^4} e^{-2\pi\sqrt{-1}\mathbf{x} \cdot \xi} \varphi(\mathbf{x}) d\mathbf{x}, \qquad \mathcal{F}^{-1}[\varphi](\mathbf{x}) = \int_{\mathbb{R}^4} e^{2\pi\sqrt{-1}\mathbf{x} \cdot \xi} \varphi(\xi) d\xi.$$

For each $\epsilon > 0$, let $j_\epsilon^{\frac{1}{2}}(\xi)$, $j^{\frac{1}{2}}(\xi)$, $j_\epsilon(\xi)$, and $j(\xi)$, resp., be the symbol of the pseudo differential operators, resp., such that

$$j_\epsilon^{\frac{1}{2}}(\xi) \equiv \left(|\xi|^2 + 1 + \epsilon(|\xi|^2 + 1)^2\right)^{-\frac{1}{2}}, \qquad j^{\frac{1}{2}}(\xi) \equiv \left(|\xi|^2 + 1\right)^{-\frac{1}{2}},$$

$$j_\epsilon(\xi) \equiv \left(|\xi|^2 + 1 + \epsilon(|\xi|^2 + 1)^2\right), \qquad j(\xi) \equiv \left(|\xi|^2 + 1\right),$$

and define

$$(J_\epsilon^{\frac{1}{2}}\varphi)(\mathbf{x}) = \mathcal{F}^{-1}(j_\epsilon^{\frac{1}{2}}\hat{\varphi})(\mathbf{x}), \qquad (J^{\frac{1}{2}}\varphi)(\mathbf{x}) = \mathcal{F}^{-1}(j^{\frac{1}{2}}\hat{\varphi})(\mathbf{x}),$$

$$(J_\epsilon\varphi)(\mathbf{x}) = \mathcal{F}^{-1}(j_\epsilon\hat{\varphi})(\mathbf{x}), \qquad (J\varphi)(\mathbf{x}) = \mathcal{F}^{-1}(j\hat{\varphi})(\mathbf{x}).$$

Symbolically

$$J_\epsilon^{\frac{1}{2}} = \left(-\Delta_{d=4} + 1 + \epsilon(-\Delta_{d=4} + 1)^2\right)^{-\frac{1}{2}}, \qquad J^{\frac{1}{2}} = (-\Delta_{d=4} + 1)^{-\frac{1}{2}},$$

$$J_\epsilon = \left(-\Delta_{d=4} + 1 + \epsilon(-\Delta_{d=4} + 1)^2\right)^{-1}, \qquad J = (-\Delta_{d=4} + 1)^{-1},$$

where

$$\Delta_{d=4} \equiv \frac{\partial}{\partial t} + \frac{\partial}{\partial x} + \frac{\partial}{\partial y} + \frac{\partial}{\partial z}, \qquad \text{with} \quad \mathbf{x} = (t, x, y, z).$$

Let

$$< \phi(\omega), \varphi >_{\mathcal{S}',\mathcal{S}} = \int_{\mathbb{R}^4}\left(\int_{\mathbb{R}^4} \varphi(\mathbf{x})J^{\frac{1}{2}}(\mathbf{x} - \mathbf{y})d\mathbf{x}\right)\dot{W}(\mathbf{y})d\mathbf{y}, \text{for } \varphi \in \mathcal{S}'(\mathbb{R}^4 \to \mathbb{R}). \tag{2.1}$$

Also, define an $\mathcal{S}'(\mathbb{R}^4 \to \mathbb{R})$-valued random variable $\phi_\epsilon(\omega)$ such that

$$< \phi_\epsilon(\omega), \varphi >_{\mathcal{S}',\mathcal{S}} = \int_{\mathbb{R}^4}\left(\int_{\mathbb{R}^4} \varphi(\mathbf{x})J_\epsilon^{\frac{1}{2}}(\mathbf{x} - \mathbf{y})d\mathbf{x}\right)\dot{W}(\mathbf{y})d\mathbf{y}, \text{for } \varphi \in \mathcal{S}'(\mathbb{R}^4 \to \mathbb{R}), \tag{2.2}$$

and for $p \in \mathbb{N}$, define

$$<: \phi_\epsilon^p :, \varphi > = \int_{(\mathbb{R}^4)^p}\left\{\int_{\mathbb{R}^4} \varphi(\mathbf{x})\prod_{i=1}^{p} J_\epsilon^{\frac{1}{2}}(\mathbf{x} - \mathbf{y}_i)d\mathbf{x}\right\} : \prod_{j=1}^{p} \dot{W}(\mathbf{y}_j) : \prod_{j=1}^{p} d\mathbf{y}_j. \tag{2.3}$$

By using the expansion by means of the *Wiener chaos* (i.e., Fock space representation), we see that $(<: \phi_\epsilon^p :, 1_\Lambda >)^k$ with $p = 2q$, $q \in \mathbb{N}$, has estimated $(k!)^q$ terms (cf. (3.7) below). We consider

$$(<: \phi_\epsilon^p :, 1_\Lambda >)^k = \sum_{G' \in A'(k;\epsilon,\Lambda)} G' + \sum_{G \in A(k;\epsilon,\Lambda)} G, \tag{2.4}$$

(cf. [FelMagRivSé], [CRiv]). Where $A'(k; \epsilon, \Lambda)$ is a set of random variables such that for $G'(k; \epsilon, \Lambda) \in A'(k; \epsilon, \Lambda)$ there exists an $n \in \mathbb{N} \cup \{0\}$ and

$$\lim_{\epsilon \downarrow 0, \, \Lambda \uparrow \mathbb{R}^4} E\left[(\phi(f_1) \cdots \phi(f_n) \, G'(k; \epsilon, \Lambda)) \right] \text{ diverges}, \forall f_j \in \mathcal{S}(\mathbb{R}^4), j = 1, \ldots, n$$

and $A(k; \epsilon, \Lambda)$ is a set of random variables such that for $G(k; \epsilon, \Lambda) \in A(k; \epsilon, \Lambda)$

$$\lim_{\epsilon \downarrow 0, \, \Lambda \uparrow \mathbb{R}^4} E\left[(\phi(f_1) \cdots \phi(f_k) \, G(k; \epsilon, \Lambda)) \right] \text{ converges},$$

$$\forall f_j \in \mathcal{S}(\mathbb{R}^4), \quad j = 1, \ldots, n, \quad \forall n \in \mathbb{N} \cup \{0\}. \tag{2.5}$$

In the next section (cf. (3.14)), for some elements in $A(k; \epsilon, \Lambda)$, we shall explicitly define *Hida distributions*:

$$G = \lim_{\epsilon \downarrow 0, \, \Lambda \uparrow \mathbb{R}^4} G(k; \epsilon, \Lambda). \tag{2.6}$$

Definition 2.1: [Tensor Product] Let F and G be the random variables in $\cap_{p \geq 1} L^p(\Omega; P)$ defined by the multiple stochastic integral with respect to $\dot{W}$ such that

$$F = \int_{(\mathbb{R}^4)^n} f(\mathbf{x}_1, \ldots, \mathbf{x}_n) : \prod_{k=1}^{n} \dot{W}(\mathbf{x}_k) : d\mathbf{x}_1 \cdots d\mathbf{x}_n, \tag{2.7}$$

$$G = \int_{(\mathbb{R}^4)^m} g(\mathbf{x}_1, \ldots, \mathbf{x}_m) : \prod_{k=1}^{m} \dot{W}(\mathbf{x}_k) : d\mathbf{x}_1 \cdots d\mathbf{x}_n. \tag{2.8}$$

The *tensor product* $F \otimes G$ of F and G is defined by

$$F \otimes G$$

$$= \int_{(\mathbb{R}^4)^{n+m}} f(\mathbf{x}_1, \ldots, \mathbf{x}_n) g(\mathbf{x}_{n+1}, \ldots, \mathbf{x}_{n+m}) : \prod_{k=1}^{n+m} \dot{W}(\mathbf{x}_k) : d\mathbf{x}_1 \cdots d\mathbf{x}_{n+m}. \tag{2.9}$$

As a usual notation by [GrotStreit], the notion expressed by $F \otimes G$ above is equivalent with the Wick product $F \diamond G$. ∎

Definition 2.2: [simplified definition of Hida distributions] For $f^{(n)}(\mathbf{x}_1, \ldots, \mathbf{x}_n) \in \mathcal{S}((\mathbb{R}^4)^n)$, $n \in \mathbb{N}$ ($f^{(0)}$ is understood as a constant), a random variable φ defined by

$$\varphi = \sum_{n=0}^{\infty} \int_{(\mathbb{R}^4)^n} f^{(n)}(\mathbf{x}_1, \ldots, \mathbf{x}_n) : \prod_{k=1}^{n} \dot{W}(\mathbf{x}_k) : d\mathbf{x}_1 \cdots d\mathbf{x}_n,$$

is said to be in $(\mathcal{S})_r$ for $r \in \mathbb{N} \cup \{0\}$ if

$$\|\varphi\|^2_{L^2(\Omega,P),r} \equiv \sum_{n=0}^{\infty} n! \left\| \left(\prod_{k=1}^{n} (-\Delta_{\mathbf{x}_k} + |\mathbf{x}_k|^2 + 1)^r \right) f^{(n)} \right\|^2_{L^2((\mathbb{R}^4)^n)} < \infty,$$

$$(2.10)$$

where $\Delta_{\mathbf{x}_k} \equiv \dfrac{\partial^2}{\partial^2 t_k} + \dfrac{\partial^2}{\partial^2 x_k} + \dfrac{\partial^2}{\partial^2 y_k} + \dfrac{\partial^2}{\partial^2 z_k}$ for $\mathbf{x}_k = (t_k, x_k, y_k, z_k) \in \mathbb{R}^4$.
$\|\varphi\|^2_{L^2(\Omega,P),r}$ is equivalent with

$$E \left[\left(\sum_{n=0}^{\infty} \int_{(\mathbb{R}^4)^n} \left(\prod_{k=1}^{n} (-\Delta_{\mathbf{x}_k} + |\mathbf{x}_k|^2 + 1)^r \right) f^{(n)}(\mathbf{x}_1, \ldots, \mathbf{x}_n) : \prod_{k=1}^{n} \dot{W}(\mathbf{x}_k) : \right. \right.$$
$$\left. \left. \times \, d\mathbf{x}_1 \cdots d\mathbf{x}_n \right)^2 \right].$$

We say that a sequence of random variables $\{G_\epsilon\}_{\epsilon>0}$ defines a Hida distribution $G \in (\mathcal{S})_{-r}$ if there exists a constant $K < \infty$ and

$$|E\left[G_\epsilon \cdot \varphi\right]| \le K \|\varphi\|_{L^2(\Omega,P),r}, \qquad \forall \varphi \in (\mathcal{S})_r, \qquad (2.11)$$

and the limit exists for $\forall \varphi \in (\mathcal{S})_r$:

$$\lim_{\epsilon \downarrow 0} E\left[G_\epsilon \, \varphi\right]. \qquad (2.12)$$

The following continuous linear functional G on $(\mathcal{S})_r$ is called as a Hida distribution in $(\mathcal{S})_{-r}$:

$$< G, \varphi > = \lim_{\epsilon \downarrow 0} E\left[G_\epsilon \, \varphi\right]. \qquad (2.13)$$

∎

Remark 2.3: i)   If the defining sequence $\{G_\epsilon\}_{\epsilon>0}$ of a *Hida distribution* G is composed by elements of *n-times* multiple stochastic integrals, then

$$E \left[G_\epsilon \cdot \left(\int_{(\mathbb{R}^4)^m} f(\mathbf{x}_1, \ldots, \mathbf{x}_m) : \prod_{k=1}^{m} \dot{W}(\mathbf{x}_k) : d\mathbf{x}_1 \cdots d\mathbf{x}_m \right) \right] = 0, \quad (2.14)$$

for any $f \in \mathcal{S}(\mathbb{R}^m)$ with $m \ne n$.

ii)   Since,

$$\left\| \left(\prod_{k=1}^{n} (-\Delta_{\mathbf{x}_k} + |\mathbf{x}_k|^2 + 1)^r \right) f \right\|_{L^2((\mathbb{R}^4)^n)}$$

$$= \left\{ \int_{(\mathbb{R}^4)^n} \left(\left(\prod_{k=1}^{n} (|\mathbf{x}_k|^2 + 1)^{-2} (|\mathbf{x}_k|^2 + 1)^2 (-\Delta_{\mathbf{x}_k} + |\mathbf{x}_k|^2 + 1)^r \right) f \right)^2 \right.$$

$$\left. \times\, d\mathbf{x}_1 \cdots d\mathbf{x}_n \right\}^{\frac{1}{2}}$$

$$\leq \left(\sup_{\mathbf{x}_1,\ldots,\mathbf{x}_n} \left| \left(\prod_{k=1}^{n} (|\mathbf{x}_k|^2 + 1)^2 (-\Delta_{\mathbf{x}_k} + |\mathbf{x}_k|^2 + 1)^r \right) f \right| \right)$$

$$\cdot \left\| \prod_{k=1}^{n} (|\mathbf{x}_k|^2 + 1)^{-2} \right\|_{L^2((\mathbb{R}^4)^n)}$$

$$\leq K \cdot p_{m,k}(f), \qquad\qquad \forall f(\mathbf{x}_1,\ldots,\mathbf{x}_n) \in \mathcal{S}((\mathbb{R}^4)^n), \quad (2.15)$$

where $p_{m,k}(\cdot)$ is a semi-norm of $\mathcal{S}((\mathbb{R}^4)^n)$ such that

$$p_{m,k}(f) = \sum_{|\alpha| \leq m} \sup_{\overline{\mathbf{x}}} (1 + |\overline{\mathbf{x}}|^2)^k |D^\alpha f(\overline{\mathbf{x}})|, \qquad\qquad (2.16)$$

with

$$\overline{\mathbf{x}} = (\mathbf{x}_1,\ldots,\mathbf{x}_n) \in (\mathbb{R}^4)^n, \qquad \mathbf{x}_k = (t_k, x_k, y_k, z_k) \in \mathbb{R}^4,$$

$$\alpha = (\alpha_1,\ldots,\alpha_n), \quad \alpha_k = (\alpha_{k1}, \alpha_{k2}, \alpha_{k3}, \alpha_{k4}), \quad |\alpha| = \sum_{k=1}^{n} \sum_{i=1}^{4} \alpha_{ki},$$

$$D^\alpha = \prod_{k=1}^{n} \left(\frac{\partial}{\partial t_k} \right)^{\alpha_{k1}} \left(\frac{\partial}{\partial x_k} \right)^{\alpha_{k2}} \left(\frac{\partial}{\partial y_k} \right)^{\alpha_{k3}} \left(\frac{\partial}{\partial z_k} \right)^{\alpha_{k4}}.$$

By this, if a *Hida distribution* G is defined through a sequence $\{G_\epsilon\}_{\epsilon>0}$ of *n-times* multiple stochastic integrals, then it can be *identified* with an element of $\mathcal{S}'((\mathbb{R}^4)^n)$:
$\exists K < \infty$ and $\exists m,\, k \in \mathbb{N}$ that depend only on r and

$$| <G, \varphi> | \leq K \cdot p_{m,k}(f)$$

$$\mathcal{S}'((\mathbb{R}^4)^n) \ni G \,:\, \mathcal{S}((\mathbb{R}^4)^n) \ni f \longmapsto\, <G, \varphi>, \qquad\qquad (2.17)$$

for

$$\varphi = \int_{(\mathbb{R}^4)^n} f(\mathbf{x}_1,\ldots,\mathbf{x}_n) : \prod_{k=1}^n \dot{W}(\mathbf{x}_k) : d\mathbf{x}_1 \cdots d\mathbf{x}_n.$$

∎

3. Well Defined Terms (Feynman Graphs) for $d = 4$ as Hida Distributions

Through the discussions in the previous sections, for $\epsilon > 0$, we can define

$$e^{-\lambda <:\phi_\epsilon^{2q}:,1_\Lambda>} = \sum_{k=0}^{\infty} \frac{(-\lambda)^k}{(k!)} \left(<: \phi_\epsilon^{2q} :, 1_\Lambda >\right)^k. \tag{3.1}$$

The equality holds for $P - a.e.\ \omega \in \Omega$, because both sides of the equality are real valued random variables, regardless they are integrable or not. The number of terms (graphs) of $\left(<: \phi_\epsilon^{2q} :, 1_\Lambda >\right)^k$, is estimated by $(k!)^q$:

$$\left(<: \phi_\epsilon^4 :, 1_\Lambda >\right)^k = \sum_{G \in \tilde{A}(k;\epsilon,\Lambda)} G, \tag{3.2}$$

where, each $G \in \tilde{A}(k; \epsilon, \Lambda)$ is a *tensor product of* multiple stochastic integrals, and in the sense of Feynman graph it is a graph with k-vertices, also the cardinality of the set $\tilde{A}(k; \epsilon, \Lambda)$ is the order $(k!)^q$. By the notation of (2.4), $\tilde{A}(k; \epsilon, \Lambda)$ can be expressed by the direct sum

$$\tilde{A}(k; \epsilon, \Lambda) - A'(k; \epsilon, \Lambda) \oplus A(k; \epsilon, \Lambda). \tag{3.3}$$

For $r \geq 0$, let

$$\Lambda_r \equiv \left\{ (t,x,y,z) \in \mathbb{R}^4 \,\Big|\, \sqrt{t^2 + x^2 + y^2 + z^2} \leq r \right\}. \tag{3.4}$$

For each $k \in \mathbb{N} \cup \{0\}$, define $A_r^\circ(\epsilon, k)$, a subset of $A(k; \epsilon, \Lambda_r) \subset \tilde{A}(k; \epsilon, \Lambda_r)$ in (3.3), as follows:

$A_r^\circ(\epsilon, k) =$ the set with the elements $G \in \tilde{A}(k; \epsilon, \Lambda_r)$ (cf. (3.14)), each of which is a tensor product of stochastic integrals, and each component of the tensor product is identified with a connected Feynman graph that is in *type O* or *type I*. (3.5)

Where, *type O* resp., and *type I* are defined as follows:
Type O is the set of graphs, each element of which satisfies the following:
(O-i)   each vertex of the graph has $2q - 2$ free (open) legs,
(O-ii)   each vertex connects with distinguished two vertices,

(O-iii)   *there exists a connected path passing through every vertex of the graph exactly once.*

Type I is the set of graphs, each element of which satisfies the following:

(I)   *between each vertex and the other vertex in the graph there exists exactly only one (directed) path that connects these two vertices, and each vertex of the graph has $2q - 2$ or $2q - 1$ free (open) legs.*

Then, **formally** (cf. [CRiv], [Sok]), by (3.1) and (3.2),

$$E\left[\phi(f_1)\cdots\phi(f_n)\, e^{-\lambda<:\phi_\epsilon^{2q}:,1_{\Lambda_r}}\right]$$

$$= E\left[\phi(f_1)\cdots\phi(f_n)\left(\sum_{k=0}^{\infty}\frac{(-\lambda)^k}{k!}\left(\sum_{G\in\tilde{A}(k;\epsilon,\Lambda_r)}G\right)\right)\right]$$

for $f_j \in \mathcal{S}(\mathbb{R}^4 \to \mathbb{R})$, $j = 1,\ldots,n$, $n \in \mathbb{N} \cup \{0\}$.

We shall show that $A_r^\circ(\epsilon,k) \subset A(k;\epsilon,\Lambda_r)\,(\subset \tilde{A}(k;\epsilon,\Lambda_r))$, namely for $G \in A_r^\circ(\epsilon,k)$,

$$\lim_{t\to\infty}\lim_{\epsilon\downarrow 0}E\left[\phi(f_1)\cdots\phi(f_n)\,G\right]$$

exists, and the limit is understood as a *Hida distribution*.

To give the statements we prepare some notions and notations:

Denote the number of the elements of $A_r^\circ(\epsilon,k)$ by $N(A(k))$, and give an index to each $G \in A_r^\circ(\epsilon,k)$ to indicate it as $G_j(k;\epsilon,r)$, $j = 1,\ldots,N(A(k))$. Then,

$$A_r^\circ(\epsilon,k) = \left\{G_j(k;\epsilon,r)\right\}_{j=1,\ldots,N(A(k))}, \qquad k \in \mathbb{N}. \qquad (3.6)$$

Recall that for $f_j \in \mathcal{S}(\mathbb{R}^4)$, $j = 1,2,\ldots,$

$$\phi(f_1)\phi(f_2) = :\phi(f_1)\phi(f_2): +E[\phi(f_1)\phi(f_2)]$$

$$\phi(f_1)\phi(f_2)\phi(f_3)\phi(f_4) = :\phi(f_1)\phi(f_2)\phi(f_3)\phi(f_4):$$
$$+ (:\phi(f_1)\phi(f_2):)\,E[\phi(f_3)\phi(f_4)] + (:\phi(f_1)\phi(f_3):)\,E[\phi(f_2)\phi(f_4)]$$
$$+ (:\phi(f_1)\phi(f_4):)\,E[\phi(f_2)\phi(f_3)] + (:\phi(f_2)\phi(f_3):)\,E[\phi(f_1)\phi(f_4)]$$
$$+ (:\phi(f_2)\phi(f_4):)\,E[\phi(f_1)\phi(f_3)] + (:\phi(f_3)\phi(f_4):)\,E[\phi(f_1)\phi(f_2)]$$
$$+ \sum_{\text{distinguished } i_l\text{'s}} E[\phi(f_{i_1})\phi(f_{i_2})]E[\phi(f_{i_3})\phi(f_{i_4})], \qquad (3.7)$$

and

$$:\phi(f_1)\cdots\phi(f_n):= \int_{(\mathbb{R}^4)^n}\prod_{j=1}^{n}(J^{\frac{1}{2}}f_j)(\mathbf{x}_j):\prod_{j=1}^{n}\dot{W}(\mathbf{x}_j):\,d\mathbf{x}_1\cdots d\mathbf{x}_n. \quad (3.8)$$

By using the above, through a simple evaluation we have the following:

Lemma 3.1: *For each $r \geq 0$, $\epsilon > 0$ and $k \geq 1$, let $A_r^\circ(\epsilon, k)$ be the set of Feynman graphs (i.e., set of multiple stochastic integrals) defined by (3.5). Then, by using the expression (3.6) and (3.7), the following hold:*

$$N(A(k)) \simeq (k!)^{\frac{3}{2}}, \tag{3.9}$$

and by denoting the number of free legs of $G_j(k; \epsilon, r) \in A_r^\circ(\epsilon; k)$ by $N_f(G_j(k; \epsilon, r))$,

$$k + 2 \leq N_f(G_j(k; \epsilon, r)) \leq 4k, \qquad j = 1, \ldots, N(A(k)), \tag{3.10}$$

also (cf. (3.7))

$$E\left[(: \phi(f_1) \cdots \phi(f_n) :) \, G_j(k; \epsilon, r)\right] = 0, \quad \text{if } n \neq N_f(G_j(k; \epsilon, r)), \tag{3.11}$$

moreover, there exists $M < \infty$ and

$$\left| E\left[(: \phi(f_1) \cdots \phi(f_n) :) \, G_j(k; \epsilon, r)\right] \right| \leq M^k (n!) \prod_{i=1}^n \left\| \hat{f}_i \right\|_{L^1(\mathbb{R}^4)}, \tag{3.12}$$

for $\forall f_i \in \mathcal{S}(\mathbb{R}^4 \to \mathbb{R})$, $i = 1, \ldots, n$, $\forall n \in \mathbb{N}$; $\forall G_j(k; \epsilon, r) \in A_r^\circ(\epsilon, k)$; $\forall k \in \mathbb{N}$, $\forall \epsilon > 0$, and $\forall r > 0$ (if $n \neq N_f(G_j(k; \epsilon, r))$, then "right-hand side of (3.12)" $= 0$).

$\square$

From Definition 2.2, Remark 2.3 and Lemma 3.1, we immediately have

Theorem 3.2: *For each $k \in \mathbb{N}$ and $j = 1 \ldots, N(A(k))$, there exists a Hida distribution $G_j(k)$, which is a Feynman graph, such that*

$$\lim_{r \to \infty} \lim_{\epsilon \downarrow 0} G_j(k; \epsilon, r) = G_j(k) \qquad \text{with} \quad N_f(G_j(k)) = N_f(G_j(k; \epsilon, r)) \tag{3.13}$$

and the set $A_r^\circ(\epsilon, k)$ converges to a set of Hida distributions $A(k)$,

$$\lim_{r \to \infty} \lim_{\epsilon \downarrow 0} A_r^\circ(\epsilon, k) = A(k) = \{G_j(k)\}_{j=1,\ldots,N(A(k))}. \tag{3.14}$$

Denote

$$\langle (: \phi(f_1) \cdots \phi(f_n) :), G_j(k) \rangle \equiv \lim_{r \to \infty} \lim_{\epsilon \downarrow 0} E\left[(: \phi(f_1) \cdots \phi(f_n) :) G_j(k; \epsilon, r)\right], \tag{3.15}$$

then there exist $M < \infty$, m, $l \in \mathbb{N}$ and

$$\left| \left\langle (: \phi(f_1) \cdots \phi(f_n) :), G_j(k) \right\rangle \right| \le M^k(n!) \prod_{i=1}^{n} \|(-\Delta_{d=4} + 1)^2 f_i\|_{L^2(\mathbb{R}^4)},$$
(3.16)

$$\left| \left\langle (: \phi(f_1) \cdots \phi(f_n) :), G_j(k) \right\rangle \right| \le M^k(n!) \prod_{i=1}^{n} p_{m,l}(f_i),$$
(3.17)

for $\forall f_i \in \mathcal{S}(\mathbb{R}^4 \to \mathbb{R})$, $i = 1, \ldots, n$, $\forall n \in \mathbb{N}$; $\forall G_j(k) \in A(k)$; $\forall k \in \mathbb{N}$ (if $n \ne N_f(G_j(k))$, then "right-hand side of (3.16) and (3.17)" $= 0$), where $p_{m,l}(f)$ is the semi-norm defined by (2.16).

$\square$

References

AFeY. Albeverio, S., Ferrario, B., Yoshida, M.W., On the essential self-adjointness of Wick powers of relativistic fields and of fields unitary equivalent to random fields, *Acta Applicande Mathematicae* **80** 309-334 (2004).

AGW1. Albeverio, S., Gottschalk, H., Wu, J.-L., Convoluted generalized white noise, Schwinger functions and their analytic continuation to Wightman functions, *Rev. Math. Phys.* **8** (1996) 763-817.

AGW2. Albeverio, S., Gottschalk, H., Wu, J.-L., Models of local relativistic quantum fields with indefinite metric (in all dimensions), *Comm. Math. Phys.* **184** (1997), 509-531.

AH-K. Albeverio, S., Høegh-Krohn, R., Uniqueness and the global Markov property for Euclidean fields: The case of trigonometric interactions, *Comm. Math. Phys.* **68** (1979), 95-128.

AR. Albeverio, S., Röckner, M., Classical Dirichlet forms on topological vector spaces-closability and a Cameron-Martin formula, *J. Functional Analysis* **88** (1990) 395-436.

AY1. Albeverio, S., Yoshida, M. W., $H - C^1$ maps and elliptic SPDEs with polynomial and exponential perturbations of Nelson's Euclidean free field, *J. Functional Analysis* **196** (2002) 265-322.

AY2. Albeverio, S., Yoshida, M. W., Hida distribution construction of non-Gaussian reflection positive generalized random fields, *Infinite Dimensional Analysis, Quantum Probability and Related Topics.* **12** (2009) 21-49.

AY3. Albeverio S., Yoshida M. W., Hida distribution construction of $P(\phi)_d$ ($d \ge 4$) indefinite metric quantum field models without BPHZ renormalization *RIMS Kyoto Kôkyûroku* (2014). (Applications of renormalizing methods in mathematical sciences. 2013. Sept.)

Araki. Araki, H., On a pathorogy in indefinite inner product spaces, *Commn. Math. Phys.* **85** (1982), 121-128.

BaSeZh. Baez, J.C., Segal, I.E., Zhou, Z., *Introduction to Algebraic and Constructive Quantum Field Theory*, Princeton Univ. Press, 1992.

BogPara. Bogoliubow, N. N., Parasiuk, O. S., *Über die Multiplikation der Kausalfunktionen in der Quantentheorie der Felder*, **97** (1957), 227-266.

BogLogT. Bogoliubov, N. N., Logunov, A. A., Todorov, I. T., *Introduction to axiomatic quantum field theory*, Translated from the Russian by Stephen A. Fulling and Ludmila G. Popova. Edited by Stephen A. Fulling. Mathematical Physics Monograph Series, No. 18. W. A. Benjamin, Inc., Reading, Mass.-London-Amsterdam, 1975.

Bo. Borchers, H.-J., *Algebraic aspects of Wightman field theory*, in R.N. Sen and C. Weil (eds.), Statistical mechanics and Field Theory, Haifa Lectures 1971; New York: Halstedt Press, 1972.

BrFSok. Brydges, D. C., Fröhlich, J., Sokal, A. D., A new proof of the existence and nontriviality of the continuum φ_2^4 and φ_3^4 quantum field theory, *Comm. Math. Phys.* **91** (1983) 141-186.

CRiv. de Calan, C., Rivasseau, V., Local existence of the Borel transform in Euclidean Φ_4^4, *Commn. Math. Phys.* **82** (1981) 69-100.

DaFelRiv. David, F., Feldman, J., Rivasseau, V., On larde order bahavior of Φ_4^4, *Commn. Math. Phys.* **116** (1988) 215-233.

DüR. Dütsch, M., Rehren, K.-H., A comment on the dual field in the AdS-CFT correspondence, Lett. Math. Phys. 62 (2002), no. 2, 171-184

Ep. Epstein, H., On the Borchers class of a free field, *Nuovo Cimento* **27** 1963, 886-893.

Fel. Feldman, J., The $\lambda \varphi_3^4$ field theory in a finite volume, *Commn. Math. Phys.* **37** (1974) 93-120.

FelMagRivSé. Feldman, J., Magnen, J., Rivasseau, V., Sénéor, R., Bound on completely convergent Euclidean Feynman graphs, *Commn. Math. Phys.* **98** (1985) 273-288.

FernFSok. Fernandez, R., Fröhlich, J., Sokal, A. D., *Random walks, critical phenomena, and triviality in quantum field theory*, Texts and Monographs in Physics. Springer-Verlag, Berlin, 1992.

GRSi. Guerra, F., Rosen, L., Simon, B., The $P(\phi)_2$ Euclidean quantum field theory as classical statistical mechanics. I, II, *Ann. of Math.* **101** (1975) 111-189

GJ. Glimm, J., Jaffe, A., *Quantum Physics: A Functional Integral Point of View*, 2nd ed., Springer, Berlin, 1987.

GrotStreit. Grothaus, M., Streit, L., Construction of relativistic quantum fields in the framework of white noise analysis, *J. Math. Phys.* **40** (1999) 5387-5405.

Heg. Hegerfeldt, G. C., From Euclidean to relativistic fields and on the notion of Markoff fields, *Comm. Math. Phys.* **35** (1974) 155-171.

Hepp. Hepp, K., Proof of the Bogoliubov-Parasiuk theorem on renormalization, *Comm. Math. Phys.* **2** (1966) 301-326.

H1. Hida, T., Generalized multiple Wiener integrals, *Proc. Japan Acad. Ser. A Math. Sci.* **54** (1978) 55–58.

H2. Hida, T., *Brownian motion*, Springer-Verlag, New York Heidelberg Berlin 1980.

HKP Streit. Hida, T., Kuo, H.-K., Potthoff, J., Streit, L., *White Noise: An Infinite Dimensional Calculus*, Kluwer Academic Publishers, Dordrecht, 1993.

HStreit. Hida, T., Streit,L, On quantum thery in terms of white noise, *Nagoya Math. J.* **68** (1977) 21-34.

Ho. Hofman, G., On GNS representations on inner product space, *Commn. Math. Phys.* **191** (1998) 299-323.

IkW. Ikeda, N., Watanabe, S., *Stochastic differential equations and diffusion processes*, second edition, North-Holland, 1989.

ItoKR. Ito, K. R., *Publ. RIMS Kyoto Univ.* **14** (1978) 503.

Jo. Jost, R., *The general theory of quantized fields*, Ed. Mark Kac, Lectures in Applied Mathematics (Proceedings of the Summer Seminar, Boulder, Colorado, 1960), Vol. IV American Mathematical Society, Providence, R.I. 1965 xv+157 pp.

Klein1. Klein, A., Renormalized products of the generalized free field & its derivatives, *Pac. J. Math.* **45** (1973) 275-292.

Klein2. Klein, A., Gaussian OS-positive processes, *Z. Wahrscheinlichkeitstheorie und Verw. Gebiete* **40** (1977) 115-124.

MagNicRivSé. Magnen, J., Nicolo, F., Rivasseau, V., Sénéor, R., A Lipatov bound for Φ_4^4 Euclidean field theory, *Comm. Math. Phys.* **108** (1987) 257-289.

Mizohata. Mizohata, S.: *The theory of partial differential equations*. Cambridge University Press, New York, 1973.

MoStro. Morchio, G., Strocchi, F., Infrared singulalities, vacuum structure and pure phase in local quantum field theory, *Ann. Inst. H. Poincaré* **A33** (1980) 251-282.

NaMu1. Nagamachi, S., Mugibayashi, N., Hyperfunction quantum field theory, *Comm. Math. Phys.* **46** (1976) 119-134.

NaMu2. Nagamachi, S., Mugibayashi, N., Hyperfunction quantum field theory. II. Euclidean Green's functions, *Comm. Math. Phys.* **49** (1976) 257-275.

NaMu3. Nagamachi, S., Mugibayashi, N., Hyperfunctions and renormalization, *J. Math. Phys.* **27** (1986) 832-839.

Ne1. Nelson, E., Construction of quantum fields from Markoff fields, *J. Functional Analysis* **12** (1973) 97–112.

Ne2. Nelson, E., The free Markov field, *J. Functional Analysis* **12** (1973) 221-227.

Nu. Nualart, D., *The Malliavin calculus and related topics*, Springer-Verlag, New York/Heidelberg/Berlin, 1995.

Oj1. Ojima, I., Entropy production and nonequilibrium stationarity in quantum dynamical systems, Physical meaning of van Hove limit., *J. Statist. Phys.* **56** (1989) 203-226.

Oj2. Ojima, I.: *How to formulate non-equilibrium local states in QFT? General characterization and extension to curved spacetime*. A garden of quanta, 365-384, World Sci. Publishing, River Edge, NJ, 2003.

OS1. Osterwalder,K.,Schrader, R., Axioms for Euclidean Green's functions I, *Comm. Math. Phys.* **31** (1973) 83-112.

OS2. Osterwalder,K.,Schrader, R., Axioms for Euclidean Green's functions II, *Comm. Math. Phys.* **42** (1975) 281-305.

ReSi. Reed, M., Simon, B., *Methods of modern mathematical physics. II. Fourier analysis, self-adjointness*, Academic Press, 1975.

Rehren. Rehren, K-H., Comments on a Recent Solution to Wightman's Axioms,

Comm. Math. Phys. **178** (1996) 453-465.

Si1. Simon, B., *The $P(\Phi)_2$ Euclidean (Quantum) Field Theory*, Princeton Univ. Press, Princeton, NJ., 1974.

Si2. Simon, B., Borel summability of the ground-state energy in spacially cutoff $(\varphi^4)_2$, *Phys. Rev. letters* **25** (1970) 1583-1586.

SmSol1. Smirnov, A.G., Soloviev, M.A., Spectral properties of Wick power series of a free field with an indefinite metric, *Theoret. and Math. Phys.* **125** (2000) 1349-1362.

SmSol2. Smirnov, A.G., Soloviev, M.A., Wick power series that converge to non-local fields, *Theoret. and Math. Phys.* **127** (2001) 632-645.

Sok. Sokal, A. D., An improvement of Watson's theorem on Borel summability, *J. Math. Phys.* **21** (1980) 261-263.

Speer. Speer, E. R., On the structure of analytic renormalization, *Comm. Math. Phys.* **23** (1971) 23-36.

StWi. Streater R.F., Wightman A.S., *PCT, Spin and Statistics, and all that*, Princeton Univ. Press, 1964.

Stro. Strocchi F., *Selected topics on the General Properties of Quantum Field Theory*, Lect. Notes in Physics, **51** World Sci., Singapore-New York-London-Hong Kong, 1993.

Widder. Widder, D. V., *The Laplace transform*, Princeton University Press, Princeton, Eighth printing, 1972.

Wi1. Wightman, A.S., *Introduction to new aspects of relativistic dynamics of quantum fields*, Cargèse Lect. Theor. Phys, Ed. M. Lévy, 171-291, Gordon and Breach New York (1967).

Wi2. Wightman, A.S., *Recent achievements of axiomatic field theory*, Sem. on Theor. Phys. Trieste 1962 Int. At. En. Ag., 11-58 Vienna (1963).

Y1. Yoshida, M.W., Construction of infinite-dimensional interacting diffusion processes through Dirichlet forms, *Probab. Theory Relat. Fields* **106** (1996) 265-297.

Y2. Yoshida, M.W., Non-linear continuous maps on abstract Wiener spaces defined on space of tempered distributions, *Bulletin of the Univ. Electro-Commun.* **12** (1999) 101-117.

A MATHEMATICAL REALIZATION OF VON NEUMANN'S MEASUREMENT SCHEME

Masanari Asano

National Institute of Technology, Tokuyama College Gakuendai, Shunan
Yamaguchi 745-8585 Japan
asano@tokuyama.ac.jp

Masanori Ohya

Department of Information Sciences, Tokyo University of Science
2641 Yamazaki, Noda, Chiba, 278-8510, Japan
ohya@rs.noda.tus.ac.jp

Yuta Yamamori

Department of Information Sciences, Tokyo University of Science
Yamasaki 2641, Noda-shi, Chiba, 278-8510 Japan

In this chapter, we give a mathematical representation of von Neumann's
view on quantum measurement process, in which, a system to be measured interacts with infinite chain of measurement apparatus [1]. For
the system, the measurement apparatus is an environment, and increase of degree of freedom in the environment is the cause of quantum
decoherence of the system. We clearly represent this essential aspect in
the background of quantum measurement by using lifting maps [3, 2].
A sequence of lifting maps realizes a unitary time evolution consistent
with von Neumann's scheme.

1. Introduction

Quantum mechanics has resolved fundamental difficulties of classical physics in describing atoms and elementary particles. In spite of this success,
quantum mechanics has not yet succeeded in clarifying a fundamental problem, namely the "measurement problem". The problem of measurement
comes from the following belief: "Let us consider the process of measu-

rement in a quantum system. If quantum theory is the ultimate theory of nature, the measurement process itself be described within the quantum theory." According to the above belief, in quantum mechanics, a state change in the measurement process should be described by using unitary operator. For this purpose, the scheme proposed by von Neumann is important [1, 2]. This scheme is summarized to the following expression:

$$\exists\, \mathcal{U}_t \; s.t. \; \mathcal{U}_t |\Phi\rangle \otimes |\Psi\rangle \xrightarrow{t \to T} \sum_k c_k |k\rangle \otimes |\Psi_k\rangle.$$

In this expression, a compound system consisting of interested system and measurement apparatus is assumed. The vectors $|\Phi\rangle \in \mathcal{H}_S$ and $|\Psi\rangle \in \mathcal{H}_M$ in the LHS represent initial states of the system and measurement apparatus. The operator $\mathcal{U}_t$ is unitary to correlate the system and the apparatus, and t is a time interval needed for the evolution by $\mathcal{U}_t$. When t is a proper interval T, the compound state is achieved to the state of the RHS. Here, $\langle k|\Phi\rangle = c_k$, $\sum_k |c_k|^2 = 1$ and $\langle \Psi_i|\Psi_j\rangle = \delta_{ij}$ are satisfied. The orthogonality $\langle \Psi_i|\Psi_j\rangle = \delta_{ij}$ is essential for an ideal measurement, because $\{|\Psi_j\rangle\}$ are to be eigenstates of an observable denoted as $A = \sum_k a_k |\Psi_k\rangle\langle \Psi_k|$. ($\{a_k\}$ are measurement values.) Through the measurement, the state of system is transited to $|k\rangle$ with probability $|c_k|^2$, and then the measurement value "a_k" is obtained. To find the unitary $\mathcal{U}_t$ is a fundamental problem in the theory of measurement.

In 1972, Emch [4] mathematically described the measurement process of Stern-Gerlach experiment from von Neumann's view. Stern-Gerlach experiment determines a spin-state of electron by measuring position of electron affected by magnetic field. In the Emch's model, the spin-states $|\uparrow\rangle$, $|\downarrow\rangle$ are interpreted as the states of system $\{|k\rangle\}$, and the wave functions of electron $|\Psi_\uparrow(x)\rangle$, $|\Psi_\downarrow(x)\rangle$ correspond to the states $\{|\Psi_k\rangle\}$. Emch defined the unitary $\mathcal{U}_t$ as a time evolution operator $\exp(-\mathrm{i}H_{int}t)$ (H_{int} is an interaction Hamiltonian) and proved that $|\Psi_\uparrow(x)\rangle$ and $|\Psi_\downarrow(x)\rangle$ are orthogonal in the limit $t \to \infty$. Also, Hepp [5] proposed a model of measurement process that a spin system interacts with a macroscopic device consisting of a large number of spins. This model suggested that the infinitely many degrees of freedom in the device is closely related to the "reduction of the wave packet".

For our study in this chapter, we use a mathematical setting that is different from their approaches. We consider a simple picture of measurement process: A measurement apparatus is *a macroscopic system consisting of a large number of small systems (microscopic apparatuses)*, and these small systems are interacted with the interested system one by one. This perspective is similar to the original von Neumann's scheme; he assumed that a quantum system interacts with infinite chain of apparatus systems. In Secs. 2.1 and 2.2, we describe such a generation of interactions by using *lifting map* [3, 2]. A lifting map is defined as a map from a state of the system on $\mathcal{S}(\mathcal{H}_S)$ to a compound state of the system and one small system on $\mathcal{S}(\mathcal{H}_S \otimes \mathcal{H}_M)$, that is, a sequence of lifting maps represents a quantum dynamics that the system interacts with the small systems in a chain.

For the system, the interacted small systems in the measurement makes the total apparatus by increasing its number of small systems in the time evolution by the lifting maps. Such increase of degree of freedom causes *quantum decoherence* of system. There are several approaches to describe quantum decoherence, and especially, the approach using Gorini-Kossakowski-Sudarshan-Lindblad (GKSL) equation [6, 7, 2] is well-known. GKSL equation is the general Markovian master equation describing nonunitary evolution of the density matrix, which is expressed as a sum of two terms: The first term contributes to the time evolution of the isolated system, and the second one represents effects from environment, so-called decohering term. Note, GKSL equation describes the continuous time evolution of system only, and the decohering term is given phenomenologically. Therefore, the increase of degree of freedom is not explained explicitly. On the other hand, a discrete time evolution by lifting maps represents a step-by-step increase of degree of freedom in the environment clearly, and a quantum decoherence is found in the evolution of reduced density operator for the system, see Theorem. 2.2.1. (In the papers of [8, 9], we discussed the relation of GKSL equation and the time evolution by lifting maps.)

The time evolution we describe has a simple structure, and a realistic one will have a more complicated structure whose mathematical representation is impossible. However, our model explains an essential aspect of decoherence seen in a measurement process, and as discussed in Sec. 2.3, it gives a basic construction of the unitary $\mathcal{U}_t$ in von Neumann's scheme.

2. Representation of Measurement Process

Interaction between an interested system and a measurement apparatus is important in a quantum measurement. In this section, we mathematically describe a creation process of interactions by employing *lifting map*, which is introduced in Sec. 2.2.

2.1. *Interaction between System and Apparatus*

Firstly, we describe the initial stage of measurement, where a system and a measurement apparatus are still not interacted. To simplify the discussion, let us consider a two-level system and describe its initial state by

$$\rho \in \mathcal{S}(\mathcal{H}_S), \ \mathcal{H}_S = \mathbb{C}^2.$$

A measurement apparatus is conventionally macroscopic compared to the system. From this view, we assume that *the measurement apparatus consists of a large number of its parts (elements)*. We call these elements "small apparatuses" and assign numbers like "*j*th small apparatus". An initial state of the *j*th small apparatus is defined on Hilbert spaces $\mathcal{H}_{M_j}$, and it is denoted by

$$\sigma_j \in \mathcal{S}(\mathcal{H}_{M_j}).$$

Here, we assume that σ_j is pure:

$$\sigma_j = |\psi_j\rangle\langle\psi_j|,$$

where $|\psi_j\rangle$ is a state vector (a normalized vector) in $\mathcal{H}_{M_j}$.

We describe the measurement process as a kind of chain reaction: a large number of small apparatuses interact with the system one by one. To represent such a process, we use the following bounded operator.

Definition 2.1.1: $V_j \in \mathcal{B}(\mathcal{H}_S, \mathcal{H}_S \otimes \mathcal{H}_{M_j})$ is an isometric operator satisfying

$$V_j|x\rangle = U_j|x\rangle \otimes |\psi_j\rangle, \tag{1}$$

for any $|x\rangle \in \mathcal{H}_S$. The operator $U_j \in \mathcal{B}(\mathcal{H}_S \otimes \mathcal{H}_{M_j})$ is unitary.

The operation by $U_j \in \mathcal{B}(\mathcal{H}_S \otimes \mathcal{H}_{M_j})$ represents that the *j*th small apparatus interacts to the system. The form of U_j will depend on the feature of measurement apparatus, that is, what aspect (state) of the system the apparatus can observe. Since the system is a two-level system, the states to be observed are usually written as "$|0_S\rangle\langle 0_S|$ and $|1_S\rangle\langle 1_S|$". Here, $\{|0_S\rangle, |1_S\rangle\}$ is a CONS (complete orthonormal system) of $\mathcal{H}_S = \mathbb{C}^2$. We propose a proper U_j with the following form:

Proposition 2.1: *For* $\{|0_S\rangle, |1_S\rangle\}$*, the unitary operator* U_j *is defined as*

$$U_j \equiv |0_S\rangle\langle 0_S| \otimes U_0^j + |1_S\rangle\langle 1_S| \otimes U_1^j,$$

where U_0^j *and* U_1^j *are some unitary operators on* $\mathcal{H}_{M_j}$.

2.2. *Lifting Map and Decoherence Process*

By using the operator V_j, we define the following map:

Definition 2.2.1: $\mathcal{E}_j^* : \mathcal{S}(\mathcal{H}_S \otimes \mathcal{H}_{M_{j-1}} \otimes \cdots \otimes \mathcal{H}_{M_1}) \to \mathcal{S}(\mathcal{H}_S \otimes \mathcal{H}_{M_j} \otimes \mathcal{H}_{M_{j-1}} \otimes \cdots \otimes \mathcal{H}_{M_1})$

$$\mathcal{E}_j^*(\cdot) \equiv \underbrace{(V_j \otimes I \otimes \cdots \otimes I)}_{j} \cdot \underbrace{(V_j^* \otimes I \otimes \cdots \otimes I)}_{j}. \tag{2}$$

We call this map *lifting map*. In the terms of lifting maps, the creation of interactions is represented as

$$\Theta^{(n)} = \mathcal{E}_n^*(\mathcal{E}_{n-1}^*(\cdots(\mathcal{E}_2^*(\mathcal{E}_1^*(\Theta^{(0)})))\cdots)), \tag{3}$$

where the density operator $\Theta^{(0)} = \rho \in \mathcal{S}(\mathcal{H}_S)$ is an initial state of the system and $\Theta^{(n)} \in \mathcal{S}(\mathcal{H}_S \otimes \mathcal{H}_{M_n} \otimes \cdots \otimes \mathcal{H}_1)$ is a compound state of the system and n small apparatuses which are interacted each other. With increasing n, the correlation between the system and the measurement apparatus becomes strong. Here, to investigate the statistical property of the system, let us consider the reduced density operator defined by

$$\rho^{(n)} \equiv Tr_{\mathcal{H}_{M_n} \otimes \cdots \otimes \mathcal{H}_{M_1}}(\Theta^{(n)}) \ (n > 0). \tag{4}$$

When $\rho^{(n)}$ is represented as a 2×2 matrix like

$$\rho^{(n)} = \begin{pmatrix} \rho_{00}(n) & \rho_{01}(n) \\ \rho_{10}(n) & \rho_{11}(n) \end{pmatrix},$$

the components $\rho_{ij}(n) = \langle i_S | \rho^{(n)} | j_S \rangle$ $(i, j = 0 \text{ or } 1)$ satisfy the following properties.

Theorem 2.2.1:

(1) The diagonal part of $\rho^{(n)}$ is constant for any n;

$$diag(\rho^{(n)}) = diag(\rho^{(0)}).$$

(2) The non-diagonal part of $\rho^{(n)}$ converges to zero in the limit $n \to \infty$;

$$\lim_{n\to\infty} \cdot \begin{pmatrix} \rho_{00}(0) & \rho_{01}(n) \\ \rho_{10}(n) & \rho_{11}(0) \end{pmatrix} = \begin{pmatrix} \rho_{00}(0) & 0 \\ 0 & \rho_{11}(0) \end{pmatrix}.$$

Proof: From the definition of $\rho^{(n)}$ and Proposition 2.1.1,

$$\rho^{(n)} = Tr_{\mathcal{H}_{M_n}}(V_n^* \rho^{(n-1)} V_n)$$
$$= \begin{pmatrix} \langle 0_S | \rho^{(n-1)} | 0_S \rangle & \lambda_n \langle 0_S | \rho^{(n-1)} | 1_S \rangle \\ \lambda_n^* \langle 1_S | \rho^{(n-1)} | 0_S \rangle & \langle 1_S | \rho^{(n-1)} | 1_S \rangle \end{pmatrix}.$$

Here $\lambda_n = Tr(U_0^{n*}\sigma_n U_1^n) = \langle\psi_n|U_0^{n*}U_1^n|\psi_n\rangle$. From this result, the relations,

$$\rho_{00}(n) = \rho_{00}(n-1), \quad \rho_{11}(n) = \rho_{11}(n-1),$$
$$\rho_{01}(n) = \lambda_n\rho_{01}(n-1), \quad \rho_{10}(n) = \lambda_n^*\rho_{10}(n-1),$$

are obtained. It is clear that the diagonal part, $\rho_{00}(n)$ and $\rho_{11}(n)$, is constant for any n. Further, from $|\lambda_n| = |\langle\psi_n|U_0^{n*}U_1^n|\psi_n\rangle| < 1$ (Cauchy-Schwarz inequality), the non-diagonal part converges to zero in the limit $n \to \infty$.

$\square$

The above theorem implies a sort of decoherence that is caused by increasing of influences from "environment". In our description, the term of environment indicates the measurement apparatus (the small apparatuses interacting to the system).

2.3. *Consistency with von Neumann's Scheme*

As mentioned in Introduction, von Neumann's scheme on a measurement process is summarized to the following expression:

$$\exists\, \mathcal{U}_t \; s.t. \; \mathcal{U}_t|\Phi\rangle \otimes |\Psi\rangle \xrightarrow{t\to T} \sum_k c_k|k\rangle \otimes |\Psi_k\rangle, \tag{5}$$

where $\langle k|\Phi\rangle = c_k$, $\sum_k |c_k|^2 = 1$ and $\langle\Psi_i|\Psi_j\rangle = \delta_{ij}$. The vectors $|\Phi\rangle \in \mathcal{H}_S$ and $|\Psi\rangle \in \mathcal{H}_M$ are initial states of the system and measurement apparatus. T implies a time interval for sufficiently ideal measurement. To find the unitary $\mathcal{U}_t$ is a fundamental problem in the theory of measurement. In the bellow, we show that our description of measurement process is consistent with von Neumann's scheme.

In Eq. (3), we represented a general measurement process as

$$\Theta^{(n)} = \mathcal{E}_n^*(\mathcal{E}_{n-1}^*(\cdots(\mathcal{E}_2^*(\mathcal{E}_1^*(\Theta^{(0)})))\cdots)),$$

by using the lifting maps $\{\mathcal{E}_j^*\}$. As seen in Definition 2.2.1, a lifting map $\mathcal{E}_j^*$ is constructed by the bounded operator V_j, and as seen in Definition 2.1.1, its transformation is specified by the unitary operation $U_j \in \mathcal{B}(\mathcal{H}_S \otimes \mathcal{H}_{M_j})$. We represent each U_j in the following form.

$$\mathcal{U}_{\triangle t_j} = U_j \otimes I \otimes \cdots \otimes I \in \mathcal{B}(\mathcal{H}_S \otimes \mathcal{H}_{M_j} \bigotimes_{k\neq j} \mathcal{H}_{M_k}),$$

where the parameter $\triangle t_j$ implies the time interval for the state change (the time evolution of state) by U_j; e.g., $\triangle t_j = \frac{t(n)}{n}$. By using $\{\mathcal{U}_{\triangle t_j}\}$, we define the following unitary operator;

$$\mathcal{U}_{t(n)} = \mathcal{U}_{\triangle t_n} \circ \mathcal{U}_{\triangle t_{n-1}} \circ \cdots \circ \mathcal{U}_{\triangle t_1}, \quad t(n) = \sum_{j=1}^{n} \triangle t_j.$$

This unitary corresponds to $\mathcal{U}_t$ in Eq. (5).

Next, let us consider the simple case that the initial state is pure; $\Theta^{(0)} = |\Phi^{(0)}\rangle\langle\Phi^{(0)}|$, where $|\Phi^{(0)}\rangle = \alpha|0_S\rangle + \beta|1_S\rangle$. Then, the state $\Theta^{(n)}$ of Eq. (3) also becomes pure; $\Theta^{(n)} = |\Phi^{(n)}\rangle\langle\Phi^{(n)}|$. From Proposition 2.1.1, the vector $|\Phi^{(n)}\rangle$ has the following form:

$$|\Phi^{(n)}\rangle = \alpha|0_S\rangle \left(\bigotimes_{j=0}^{n-1} U_0^{n-j} |\psi_{n-j}\rangle \right) + \beta|1_S\rangle \left(\bigotimes_{j=0}^{n-1} U_1^{n-j} |\psi_{n-j}\rangle \right)$$

$$\equiv \alpha|0_S\rangle \otimes |\Psi_0^{(n)}\rangle + \beta|1_S\rangle \otimes |\Psi_1^{(n)}\rangle. \tag{6}$$

$|\Psi_0^{(n)}\rangle = \bigotimes_{j=0}^{n-1} U_0^{n-j}|\psi_{n-j}\rangle$ and $|\Psi_1^{(n)}\rangle = \bigotimes_{j=0}^{n-1} U_1^{n-j}|\psi_{n-j}\rangle$ are state vectors of the measurement apparatus consisting of n small apparatuses. The form of Eq. (6) naturally corresponds to RHS of Eq. (5).

As mentioned above, in von Neumann's scheme, the orthogonality of $|\Psi_0\rangle = w - \lim_{n\to\infty} |\Psi_0^{(n)}\rangle$ and $|\Psi_1\rangle = w - \lim_{n\to\infty} |\Psi_1^{(n)}\rangle$ is the important condition for ideal measurement. Let us consider the reduced density operator $\rho^{(n)} = Tr_{\mathcal{H}_{M_n} \otimes \cdots \otimes \mathcal{H}_{M_1}}(\Theta^{(n)})$ for the state of Eq. (6). It is represented as

$$\rho^{(n)} = \begin{pmatrix} |\alpha|^2 \langle\Psi_0^{(n)}|\Psi_0^{(n)}\rangle & \alpha\beta^* \langle\Psi_1^{(n)}|\Psi_0^{(n)}\rangle \\ \alpha^*\beta \langle\Psi_0^{(n)}|\Psi_1^{(n)}\rangle & |\beta|^2 \langle\Psi_1^{(n)}|\Psi_1^{(n)}\rangle \end{pmatrix}.$$

According to Theorem 2.2.1, the non-diagonal part of $\rho^{(n)}$ converges to zero in the limit $n \to \infty$, and the following result is derived.

$$\langle\Psi_k||\Psi_l\rangle = \lim_{n\to\infty} \langle\Psi_k^{(n)}|\Psi_\ell^{(n)}\rangle = \delta_{k\ell}.$$

This result suggests that if n is enough large and a proper observable is selected, an ideal measurement might be approximately realized. It is important that the number of small apparatuses n is proportional to the size of measurement apparatus and the time interval $t = t(n)$. In fact, a realistic measurement are in the finite systems and finite time, finite steps n. In such a case, we need more subtle estimation, statistically and probabilistically, which will be discussed elsewhere.

References

1. J. von Neumann, Mathematical Foundations of Quantum Mechanics, Princeton University Press, (1932/1955).
2. M. Ohya and I. Volovich, "Mathematical foudation of quantum information and computation and its applications to Nano- and Bio-Systems", Springer-Verlag, 2011.
3. L. Accardi and M. Ohya, Compound channels, transition expectations and liftings, *Applied Mathematics Optimization* **39**, 33-59 (1999), Volterra preprint N. 75 (1991).
4. G. C. Emch, On quantum measuring processes, *Helv. Phys. Acta* **45**, 1049 (1972).
5. K. Hepp, Quantum theory of measurement and macroscopic observables, *Helv. Phys. Acta* **45**, 237 (1972).
6. V. Gorini, A. Kossakowski and E. C. G. Sudarshan, Completely positive semigroups of N-level systems, *J. Math. Phys.* **17** (5), 821 (1976).
7. G. Lindblad, On the generators of quantum dynamical semigroups, *Commun. Math. Phys.* **48** (2), 119 (1976).
8. M. Asano, M. Ohya and A. Khrennikov, Quantum-like model for decision making process in two players game, *Foundations of Physics* **41** (3), 538-548 (2010).
9. M. Asano, M. Ohya, Y. Tanaka, A. Khrennikov and I. Basieva, On application of Gorini-Kossakowski-Sudarshan-Lindblad equation in cognitive psychology, *Open Systems & Information Dynamics*: **18** (1), 55-69 (2011).

ON RANDOM WHITE NOISE PROCESSES WITH MEMORY FOR TIME SERIES ANALYSIS

Christopher C. Bernido* and M. Victoria Carpio-Bernido

Research Center for Theoretical Physics, Central Visayan Institute Foundation
Jagna, Bohol, The Philippines
* cbernido.cvif@gmail.com

An analytical framework is presented for investigating time series as a stochastic process with memory. A suitable probability density function with different memory properties is evaluated using white noise analysis. This yields a memory-dependent mean square displacement which could be compared with real-world time series.

1. Introduction

A diverse collection of data now exists from fields as disparate as neuroscience, microrheology, industry, finance, telecommunications, meteorology, and other complex systems. There is a need to understand more deeply much of these data which normally are in the form of observations taken at regular time intervals at different scales, ranging from picoseconds to year-long intervals. Although much of the available data appear to be random fluctuations in time, a deeper investigation often reveals patterns or repetitions manifesting as a form of memory for the system under investigation [1, 2, 3]. In this chapter, we model random processes with memory with a variable $x(t)$ that could represent fluctuating positions of a typhoon, the jittery motion of a microparticle immersed in visco elastic material, or the irregular firing of neurons, among many possibilities. Parametrizing $x(t)$ in terms of the Gaussian white noise random variable $\omega(t)$, the probability density function can be evaluated utilizing the white noise path integral approach of T. Hida and L. Streit [4, 5]. From this, the mean square displacement (MSD) is calculated for comparison with actual time series that are often presented in terms of a time-dependent MSD.

2. Random Variables with Memory

We model a fluctuating variable with memory by parametrizing its evolution in time as [6, 7],

$$x(\tau) = x_0 + \int_0^\tau f(\tau - t) \; h(t) \; w(t) \, dt \; , \tag{1}$$

where x_0 is the initial value and $w(t)$ is the Gaussian white noise variable. As time t ranges from 0 to τ in Eq. (1), the $f(\tau - t) \; h(t)$ modulates the evolution of the white noise variable $w(t)$ thereby affecting the value or history of $x(\tau)$. The $f(\tau - t)$, in particular, can be viewed as a memory function and its explicit form, as well as that of $h(t)$, may be chosen to suit the phenomenon being modeled.

As written, Eq. (1) generates an ensemble of fluctuating paths all starting at $x(0) = x_0$, but ending in many possible values of $x(\tau)$. Suppose we would wish to pin the endpoint of the fluctuating variable $x(\tau)$ to a specific point $x(T) = x_T$, at some later time $\tau = T$. Pinning the endpoint of Eq. (1) leads us to consider only those paths which satisfy a δ-function constraint of the form,

$$\delta(x(T) - x_T) = \delta\left(x_0 + \int_0^T f(T - t) \; h(t) \; w(t) \, dt - x_T\right). \tag{2}$$

What would then be the transition probability from x_0 to x_T? The probability density function $P(x_T, T; x_0, 0)$ for fluctuations with memory can be evaluated following Feynman's sum-over-all possible histories [6, 8, 9] by evaluating the expectation value $E(\delta(x(T) - x_T))$, i.e.,

$$\begin{aligned}
P(x_T, T; x_0, 0) &= E(\delta(x(T) - x_T)) \\
&= \int \delta(x(T) - x_T) \, d\mu \\
&= \int \delta\left(x_0 + \int_0^T f(T - t) \; h(t) \; w(t) \, dt - x_T\right) d\mu,
\end{aligned} \tag{3}$$

where $d\mu$ is the Gaussian white noise measure [5].

Writing the delta function in Eq. (3) in terms of its Fourier representation we have,

$$P\left(x_T, T; x_0, 0\right) = \frac{1}{2\pi} \int d\mu \int\limits_{-\infty}^{+\infty} dk$$

$$\times \exp\left\{ik\left[\left(x_0 - x_T + \int\limits_0^T f\left(T - t\right) \, h\left(t\right) \, w\left(t\right) dt\right)\right]\right\}$$

$$= \frac{1}{2\pi} \int\limits_{-\infty}^{+\infty} dk \exp\left\{ik\left[\left(x_0 - x_T\right)\right]\right\}$$

$$\times \int \exp\left\{ik \int\limits_0^T f\left(T - t\right) h\left(t\right) w\left(t\right) dt\right\} d\mu \, . \tag{4}$$

If we let, $\xi\left(t\right) = k \, f\left(T - t\right) \, h\left(t\right)$, the integration over $d\mu$ can be carried out using [5],

$$\int \exp\left\{i \int\limits_0^T w\left(t\right) \xi\left(t\right) \, dt\right\} d\mu = \exp\left\{-\frac{1}{2} \int\limits_0^T \xi^2\left(t\right) \, dt\right\} \, . \tag{5}$$

With Eq. (5), we can then write Eq. (4) as,

$$P\left(x_T, T; x_0, 0\right) = \int\limits_{-\infty}^{+\infty} \frac{dk}{2\pi} \exp\left\{ik\left[\left(x_0 - x_T\right)\right] - \frac{k^2}{2} \int\limits_0^T \left[f\left(T - t\right) h\left(t\right)\right]^2 dt\right\} \, . \tag{6}$$

The remaining integral over dk is a Gaussian integral which can be evaluated to yield the probability density function,

$$P\left(x_T, T; x_0, 0\right) = \left(2\pi \int\limits_0^T \left[f\left(T - t\right) \, h\left(t\right)\right]^2 dt\right)^{-\frac{1}{2}}$$

$$\times \exp\left(-\left[\int\limits_0^T \left[f\left(T - t\right) \, h\left(t\right)\right]^2 dt\right]^{-1} \frac{\left(x_T - x_0\right)^2}{2}\right) \, . \tag{7}$$

There are many possible choices for the combination of $f\left(T - t\right)$ and $h\left(t\right)$ which allow an exact solution for the probability density function [7]. Note that, for the special case where $f(T-t)$ is given by a constant, $f = \sqrt{2D}$, where D is a diffusion coefficient and $h(t) = 1$, Eq. (7) is simply the Gaussian distribution,

$$P(x_T, T; x_0, 0) = \frac{1}{\sqrt{4\pi DT}} \exp\left(\frac{-(x_T - x_0)^2}{4DT}\right),$$ (8)

for the Wiener process.

3. Mean Square Displacement with Memory

From actual time series, the mean square displacement (MSD) of data points can easily be computed. The resulting time development of an empirical MSD may then be compared and contrasted with theoretical values of a stochastic process. In view of this, we now proceed to obtain an analytical expression of the MSD using Eq. (7). Designating $\langle x \rangle$ as the mean value of a fluctuating variable x, we have,

$$\text{MSD} = \left\langle (x - \langle x \rangle)^2 \right\rangle$$
$$= \langle x^2 \rangle - \langle x \rangle^2.$$ (9)

With Eq. (7), we can calculate $\langle x^2 \rangle$ as,

$$\langle x^2 \rangle = \int\limits_{-\infty}^{+\infty} x^2 \, P(x, T; x_0, 0) \, dx$$

$$= \left(2\pi \int\limits_0^T [f(T-t)h(t)]^2 \, dt\right)^{-\frac{1}{2}}$$

$$\times \int\limits_{-\infty}^{+\infty} x^2 \exp\left(-\left[\int\limits_0^T [f(T-t)h(t)]^2 \, dt\right]^{-1} \frac{(x-x_0)^2}{2}\right) dx,$$ (10)

which gives,

$$\langle x^2 \rangle = x_0^2 + \int\limits_0^T [f(T-t)h(t)]^2 \, dt.$$ (11)

With this, Eq. (9) becomes (let, $\langle x \rangle = x_0$),

$$\text{MSD} = \int\limits_0^T [f(T-t)h(t)]^2 \, dt.$$ (12)

We look at particular examples.

3.1. *Wiener process*

For a memory function $f(T - t)$ given by a constant $\sqrt{2D}$ and $h(t) = 1$, Eq. (12) yields the mean square displacement MSD_B of the Wiener process, or Brownian motion, i.e.,

$$\mathrm{MSD}_B = 2DT, \tag{13}$$

where D is the diffusion coefficient and T is time.

3.2. *Fractional Brownian motion*

Choosing a memory function of the form,

$$f(T - t) = \frac{(T - t)^{H-1/2}}{\Gamma(H + 1/2)}, \tag{14}$$

with $h(t) = 1$, Eq. (1) can be written as,

$$x(\tau) = x_0 + B^H(\tau), \tag{15}$$

where $B^H(\tau)$ is a fractional Brownian motion defined in the Riemann-Liouville representation by [10],

$$B^H(\tau) = \frac{1}{\Gamma\left(H + \frac{1}{2}\right)} \int_0^\tau (\tau - t)^{H-1/2} \, dB(t). \tag{16}$$

In Eq. (16), H is the Hurst exponent with values [10, 11, 12] $0 < H < 1$. Given Eq. (14), the mean square displacement, Eq. (12), becomes,

$$\mathrm{MSD}_{fBm} = AT^\alpha, \tag{17}$$

where T is time, $\alpha = 2H$, and $A = 1/2H \, \Gamma(H + 1/2)^2$. For $H = \frac{1}{2}$, we note that Eqs. (16) and (17) for fractional Brownian motion reduce to that of ordinary Brownian motion.

3.3. *Exponentially-modified Brownian motion*

Let us now consider a memory function $f(T - t)$ given by,

$$f(T - t) = (T - t)^{(\mu-1)/2}. \tag{18}$$

Here, we consider two possible forms of the time-dependent function $h(t)$ which could be combined with Eq. (18).

(a) One type of multiplying factor $h(t)$ is given by,

$$h(t) = t^{(\mu-1)/2} e^{-\nu t/2}. \tag{19}$$

Together with Eq. (18) the mean square displacement, Eq. (12), for this case becomes,

$$\mathrm{MSD}_{eB1} = \int_0^T (T-t)^{\mu-1} t^{\mu-1} e^{-\nu t} dt$$

$$= \sqrt{\pi}\,\Gamma(\mu)\; I_{\mu-\frac{1}{2}}(\nu T/2) \left(\frac{T}{\nu}\right)^{\mu-\frac{1}{2}} \exp(-\nu T/2), \tag{20}$$

where we used Eq. (3.388.1) of [13] for $\mathrm{Re}\,\mu > 0$, $\Gamma(\mu)$ the gamma function, and $I_\lambda(z)$ is the modified Bessel function of the first kind.

(b) Another type of multiplying factor $h(t)$ is slightly different from Eq. (19). It has the form,

$$h(t) = e^{-\beta/2t} t^{-\mu}. \tag{21}$$

Combined with Eq. (18), the mean square displacement, Eq. (12), would be,

$$\mathrm{MSD}_{eB2} = \int_0^T (T-t)^{\mu-1} e^{-\beta/t} t^{-2\mu}\, dt$$

$$= \frac{1}{\sqrt{\pi T}} \beta^{\frac{1}{2}-\mu} e^{-\frac{\beta}{2T}} \Gamma(\mu)\, K_{\mu-\frac{1}{2}}\left(\frac{\beta}{2T}\right), \tag{22}$$

where we used Eq. (3.471.4) of [13], for $\mathrm{Re}\,\mu > 0$ and $T > 0$. In Eq. (22), $K_\nu(z)$ is a modified Bessel function of the second kind.

4. Comparison with Empirical Data

4.1. Vortex track fluctuations

The movement of isolated vortices has been investigated by observing the mean square displacement of an isolated vortex forming in half a soap bubble heated at the equator [14, 15]. Realizing the resemblance of vortices wandering around the bubble with actual cyclones, this was later extended to a study of cyclone track forecast cones [15].

In [15], the actual location of cyclones or typhoons along the longitude and latitude were plotted against time to analyse fluctuations from the mean track. The log-log graph of the empirically based MSD versus time

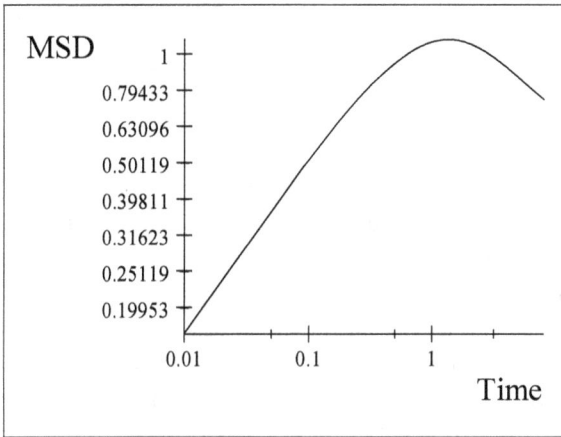

Fig. 1.   Log-log plot of MSD versus time for an exponentially modified Brownian motion given by Eqs. (18) and (19). The graph captures the general shape of MSD versus time graphs for vortices such as actual typhoons. Graphs for typhoons are shown in Figs. 3a and 3b of [15].

then showed an initially rising straight line followed by a downward curve at longer times for almost all cyclones investigated. This is exemplified by nine cyclones shown in Figures 3a and 3b of [15]. As recently reported [16], however, these empirically based MSD graphs for typhoons can be nicely modelled by a memory function $f(T - t)$ and $h(t)$. For example, using Eqs. (18) and (19), a log-log graph of MSD_{eB1} versus time T using Eq. (20) can be plotted for $\mu = 3/4$ and $\nu = 1$, as shown in Fig. 1. The downward curve appearing at longer times in Fig. 1 is similar to graphs for vortices formed in half a soap bubble, as well as graphs for real typhoons [15].

4.2. *Particle-tracking in microrheology*

Measurements of rheological properties of complex fluids such as viscosity are often done by tracking the trajectory of a micrometer size particle suspended in the medium. The medium could be colloidal suspensions, soft biological samples such as the cytoplasm of living cells, polymer solutions and gels, among many others. A tracer particle which freely diffuses normally indicates that the medium is purely viscous, while a subdiffusive path evolution indicates viscoelastic properties [17, 18, 19]. Moreover, by studying the motion of the particle when the medium is subjected to stress, insight

into the structural and mechanical response of various materials are obtained. When the applied stress is removed, a viscoelastic material returns to
its original form as a function of elapsed time. This return to its original
form after deformation, in fact, signals memory properties of viscoelastic
materials.

In various microrheological experiments, the mean square displacement
$\left\langle (\Delta r)^2 \right\rangle$ of a particle is measured at regular intervals of time. Fluctuations
of probe particles in fluids, however, seem to follow a common pattern as
shown in experimental plots of MSD versus time. Here, we model this by
considering, for example, the x-coordinate of a probe particle parametrized
as Eq. (1) where a memory function $f(T-t)$ may record the drag as t varies
from 0 to T. In particular, a combination of memory function $f(T-t)$
and $h(t)$ given by Eqs. (18) and (21), respectively, for an exponentially
modified Brownian motion appears appropriate for describing the observed
MSD of probe particles in complex fluids [17, 18, 19]. The combination yields
Eq. (22) whose log-log graph of MSD versus time t can be plotted for $\mu = \frac{1}{2}$
and $\beta = 50$, as shown in Fig. 2.

The graph given by Fig. 2, generally describes the shape of MSD versus time graphs of particle-tracking data from microrheological experiments which could provide additional insight into the nature of viscoelastic
materials. Similarities, for example, are exhibited in Figure 1 of [17] for

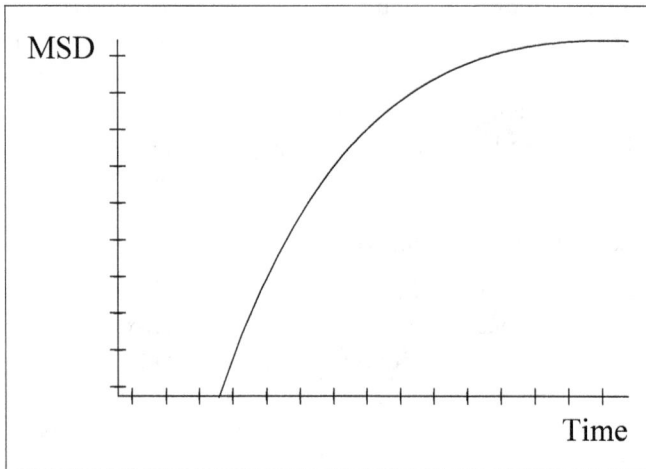

Fig. 2.   Log-log plot of MSD versus time for $\mu = 1/2$ and $\beta = 50$.

concentrated monodisperse emulsion droplets of radius $a = 0.53\mu m$, and Figure 4 of [18] for an optically trapped $4.74\mu m$ diameter silica bead suspended in water. Figure 1 (inset) of [19] for hard spheres in complex fluids also shows a similarly curved graph of MSD versus time. Fine tuning of the graph such as Fig. 2 for specific cases may be done by varying the values of μ and β in Eq. (22), or even by rescaling time to t/t_c where t_c is a constant.

5. Conclusion

A framework for investigating time series is presented in this chapter. By matching theoretical and empirical time dependence of the MSD, one would be able to determine the accuracy and appropriateness of a stochastic model for the phenomenon under investigation. Here, we cite as examples vortex track fluctuations such as those seen in typhoons and particle tracking in microrheology. From an analytical model, predictions may then be made on correlations, memory behavior and probable future values of a time series. An advantage of this approach is the wide array of possible memory functions [7] to choose from to suit the diversity of observable time series data.

Acknowledgments

The authors take this opportunity to thank Takeyuki Hida for opening up the area of white noise analysis. The support of the Institute for Mathematical Sciences, National University of Singapore, during the IDAQP workshop is also gratefully acknowledged, as well as the Alexander von Humboldt Foundation and Ludwig Streit for their visits to Universität Bielefeld.

References

1. T. Karagiannis, M. Molle, and M. Faloutsos, Long-range dependence - ten years of internet traffic modeling, *IEEE Internet Comp.*, v.5/8 (2004).
2. T. Yoshida, Universal dependence of the mean square displacement in equilibrium point vortex systems without boundary conditions, *Jour. Phys. Soc. Japan* **78** (2009) 024004.
3. P. M. Robinson, ed., *Time Series with Long Memory* (Oxford Univ. Press, 2003).
4. L. Streit and T. Hida, Generalized Brownian functionals and the Feynman integral, *Stoch. Proc. Appl.* **16** (1983) 55-69.
5. T. Hida, H. H. Kuo, J. Potthoff, L. Streit, *White Noise. An Infinite Dimensional Calculus* (Kluwer, Dordrecht, 1993).

6. C. C. Bernido and M. V. Carpio-Bernido, White noise analysis: some applications in complex systems, biophysics and quantum mechanics, *Int. Jour. Mod. Phys. B* **26** (2012) 1230014.

7. C. C. Bernido and M. V. Carpio-Bernido, *Methods and Applications of White Noise Analysis in Interdisciplinary Sciences* (World Scientific, Singapore, 2014).

8. R. P. Feynman and A. R. Hibbs, *Quantum Mechanics and Path Integrals* (McGraw-Hill, New York, 1965).

9. I. Calvo and R. Sánchez, The path integral formulation of fractional Brownian motion for the general Hurst exponent, *J. Phys. A: Math. Theor.* **41** (2008) 282002.

10. Y. Mishura, *Stochastic Calculus for Fractional Brownian Motion and Related Processes*, (Springer-Verlag, Berlin, 2008).

11. J. Klafter, S. C. Lim, R. Metzler, eds., *Fractional Dynamics* (World Scientific, Singapore, 2012).

12. F. Biagini, Y. Hu, B. Øksendal and T. Zhang, *Stochastic Calculus for Fractional Brownian Motion and Applications* (Springer-Verlag, London, 2008).

13. I. S. Gradshteyn and I. M. Ryzhik, *Table of Integrals, Series and Products*, 5th edition (Academic Press, San Diego, 1994).

14. F. Seychelles, Y. Amarouchene, M. Bessafi, and H. Kellay, Thermal convection and emergence of isolated vortices in soap bubbles. *Phys. Rev. Lett.* **100** (2008) 144501.

15. T. Meuel, G. Prado, F. Seychelles, M. Bessafi and H. Kellay, Hurricane track forecast cones from fluctuations, *Sci. Rep.* **2** (2012) 446; DOI:10.1038/srep00446.

16. C. C. Bernido, M. V. Carpio-Bernido, and M. G. O. Escobido, Modified diffusion with memory for cyclone track fluctuations, *Phys. Lett. A* **378** (2014) 2016-2019; see, also, Erratum.

17. T. G. Mason, Estimating the viscoelastic moduli of complex fluids using the generalized Stokes-Einstein equation, *Rheol. Acta* **39** (2000) 371-378.

18. M. Tassieri, R. M. L. Evans, R. L. Warren, N. J. Bailey, and J. M. Cooper, Microrheology with optical tweezers: data analysis, *New Jour. Phys.* **14** (2012) 115032.

19. T. G. Mason and D. A. Weitz, Optical measurements of frequency-dependent linear viscoelastic moduli of complex fluids, *Phys. Rev. Lett.* **74** (1995) 1250-1253.

SELF-REPELLING (FRACTIONAL) BROWNIAN MOTION RESULTS AND OPEN QUESTIONS

Jinky Bornales

*Physics Department, Mindanao State University-Iligan Institute of Technology
Iligan City, The Philippines
jinky.bornales@g.msuiit.edu.ph*

Ludwig Streit

*BiBoS, Universität Bielefeld, Germany and
CCM, Universidade da Madeira, Portugal
streit@uma.pt*

Modification of the Wiener measure by a probability density which suppresses self-intersections using the self-intersection local time was proposed by Edwards in 1965. This "polymer measure" has been under investigation ever since, within stochastic analysis, but also in the physics and chemistry communities because of the applications in the theory of macromolecules. There are existence theorems, but conjectures concerning the scaling properties of the corresponding processes have only been demonstrated rigorously in one dimension, their proof for paths in higher dimensional spaces remains an important mathematical challenge.

1. Introduction

Self-avoiding random walks are not only a classical subject of probability theory (see e.g. [24]), they are as well intensively studied in physical chemistry, in statistical and computational physics, for their important role as models for chain polymers in solvents. The interdisciplinary interplay is very stimulating — physics provides structural intuition and far-reaching predictions, and computer simulations can check them out, while the mathematical results are less far-reaching but have the higher reliability characteristic of the mathematical approach.

2. The Edwards Model

In particular insofar as their asymptotic scaling properties are concerned, self-avoiding random walks are expected to be in the same "universality class" as "weakly self-avoiding" or "self-repelling" ones where self-intersections are not excluded but suppressed by a probabilistic weight factor, such as in the Domb-Joyce model [13]. A continuum version of this is the Edwards model, informally given by the Wiener measure $d\mu_0$ with an exponential weight factor

$$d\mu_g(x) = \frac{1}{Z} \exp\left(-g \int_0^l ds \int_0^l dt\delta\left(x(s) - x(t)\right)\right) d\mu_0(x) \qquad (1)$$

with

$$Z = \mathbb{E}\left(\exp\left(-g \int_0^l ds \int_0^l dt\delta\left(x(s) - x(t)\right)\right)\right),$$

where δ is the Dirac distribution.

It is well-known that the self-intersection local time

$$L = \int_0^l ds \int_0^l dt\delta\left(x(s) - x(t)\right)$$

becomes increasingly singular for higher-dimensional Brownian motion, see e.g. [8] but also [15], and the extensive literature cited in these papers. In particular a proper definition has to pass through a regularization of the Dirac distribution, renormalizations, and limit taking. A common regularization is by the replacements

$$\delta_\varepsilon(x) := \frac{1}{(2\pi\varepsilon)^{d/2}} e^{-\frac{|x|^2}{2\varepsilon}}, \quad \varepsilon > 0,$$

$$L_\varepsilon(T) := \int_0^T dt \int_0^T ds\, \delta_\varepsilon(x(t) - x(s)). \qquad (2)$$

For an alternate "gap regularization" which omits the singular region $|t - s| < \varepsilon$ see e.g. [8].

For $d = 2$ one finds

$$\mathbb{E}(L_\varepsilon) = \frac{l}{2\pi} \ln\frac{1}{\varepsilon} + o\left(\varepsilon\right)$$

and L^2 convergence after centering. Varadhan [31] showed for

$$L_\varepsilon^c = L_\varepsilon - \mathbb{E}(L_\varepsilon)$$

and

$$L_0^c = \lim_{\varepsilon \searrow 0} L_\varepsilon^c$$

that

$$\mathbb{E}\left(\exp(-gL_0^c)\right) < \infty$$

if $g > 0$ is sufficiently small. (By a further argument this can be extended to all $g > 0$.)

3. Dirichlet Forms

In the method of "stochastic quantization", probability measures are obtained as equilibrium distributions of stochastic processes, see e.g. [1] and references there. Conversely, in [2], Álbeverio *et al.* show that in two dimensions a gradient type pre-Dirichlet form

$$\varepsilon(F) = \int |\nabla F|^2 \, d\mu_g$$

in $L^2(d\mu_g)$ is indeed closable, i.e. it generates a diffusion process with the polymer measure $d\mu_g$ as invariant distribution. They also show its ergodicity, the Dirichlet form is irreducible.

4. Generalization to Fractional Brownian Motion

Recently the Edwards model was extended to fractional Brownian motion paths

$$x(t) = B^H(t)$$

for certain values of dimension d an the Hurst index H. I.e. in formula (1) the measure μ_0 is now that of a centered Gaussian process with covariance

$$\mathbb{E}(B_t^H B_s^H) = \frac{\delta_{ij}}{2}\left(t^{2H} + s^{2H} - |t-s|^{2H}\right), i, j = 1, \ldots, d, \ s, l \geq 0, 0 < H < 1.$$

$H = 1/2$ is ordinary Brownian motion. For larger resp. smaller H the paths are smoother resp. curlier than those of Brownian motion, and continuous.

In the present context of self-intersections it is important to note a result by Talagrand [30]. He shows that, with probability 1, d-dimensional fBm paths have no $(k+1)$-tuple points whenever

$$Hd \geq \frac{k+1}{k}.$$

Existence of the fBm model has been shown as follows:

Theorem 4.1: [19] *(i) For dH = 1, d ≥ 2 there exists an M > 0 such that for all 0 ≤ g ≤ M*

$$\exp(-gL^c) \tag{3}$$

is an integrable function.
(ii) Fort dH < 1, d ≥ 2 there exists

$$L := \lim_{\varepsilon \searrow 0} L_\varepsilon \ \text{in} \ L^2$$

and for all g > 0

$$\exp(-gL)$$

is an integrable function.

5. Scaling of Self-Repelling Brownian Motion

It is intuitively clear that on average the self-repulsion, or "excluded volume" in physics and chemistry terminology, will make the trajectories of the Edwards model swell in comparison to the Brownian ones. One expects a superdiffusive behavior

$$\mathbb{E}\left(x(t)^2\right) \sim t^{2\nu} \ \text{with} \ \nu > 1/2.$$

Determining this "Flory index" $\nu = \nu_d$ for various dimensions d of the process is considered as one of the outstanding problems of probability theory. Up to now it is solved only for $d = 1$, with [20, 21]

$$\nu_1 = 1.$$

This agrees with the intuition from strictly self-avoiding random walks which can only progress linearly in one dimension, but is not at all evident in the self-repelling setting.

6. Open Questions

As mentioned above, many open questions arose and still arise from within the physics and chemistry community. To some extent any result obtained there turns into an open question for the mathematician. Hence in this section we shall use "derived", "shown" etc. to mean derived or shown to the satisfaction of the applied sciences, and will reserve the term "proof" for occasional reference to rigorous mathematical results. Another general remark concerning discretized models is that they are expected to share scaling properties of the continuum ones but are not beset with the singularity problems stemming from exact coincidence as in the Edwards model.

6.1. *The Flory index*

The Edwards model and its discretized versions are an important tool to describe the conformations of chain polymers in solvents, and the Flory scaling index was first explored in this context; its knowledge is central to the understanding of those polymers, see e.g. [17, 18, 29]. Hence it is not surprising that considerable effort has gone into its — theoretical as well as computational — determination, beginning with Flory's Nobel prize winning formula

$$\nu_d = \frac{3}{d+2} \tag{4}$$

for $d = 1, 2, 3$. Note that for $d = 4$ this would predict diffusive scaling, in keeping with the fact that there are no more self-intersections of Brownian paths for $d > 3$.

As mentioned above, the case $d = 1$ is under control. For $d = 2$, the value of $\nu_2 = 3/4$ has been confirmed using arguments from field theory and conformal invariance, as well as numerically. A rigorous proof does not exist.

The situation is similar if more pronounced, in $d = 3$. With Flory predicting $\nu_3 = 0.6$, there is agreement from renormalization group methods, computer simulations, and in fact experiment that

$$\nu_3 = 0.588,$$

for an overview see Ch. 9 of [26]. But there is no indication of a mathematical proof.

For the fractional case, in dimension d, with Hurst index H, there is a recent argument [6] that the scaling index should be

$$\nu_d^H = \frac{2H+2}{k+2}. \tag{5}$$

Since this has been derived in analogy to the simplest demonstrations the standard Flory index (4) and calls for validation. One notes that $\nu_d^H = H$ when $Hd = 2$, i.e. there is no more anomalous scaling when, as proven by Talagrand [30], there are no more self-intersections of fractional paths. There is also, following Kosmas and Freed [22] a recursion relation [6] for the scaling index and one verifies that it is fulfilled by (5). Finally, preliminary numerical tests are not in contradiction with this result.

6.2. *The Fisher-Redner-des Cloizeaux "Theorem"*

Beyond the average of the end-to-end distance, its probability distribution turns out to be of considerable interest, and is readily available in computer simulations.

It was argued first by Fisher, and then confirmed with computations and stronger arguments by Redner [28] and des Cloizeaux [10], that in dimensions $d = 2, 3$ the tail behavior of the probability distribution for the (suitably normalized) "end-to-end distance" $r = \left|x^2(t)\right|^{1/2}$ is asymptotically of the form

$$P_t(r) \sim t^{c_1} f\left(\frac{r}{t^v}\right)$$

with

$$f(x) \approx \exp\left(-c_2 x^\delta\right)$$

and, remarkably

$$\delta = \frac{1}{1 - \nu_d}.$$

In other words, the scaling index can be read off from the tail behavior of the probability distribution for sufficiently long paths or "polymers". Apart from being another mathematical challenge, this result is of considerable computational importance for the determination of the scaling behavior: instead of obtaining it from the simulation of longer and longer paths it can be computed from just one sufficiently long one, a method employed successfully e.g. by Besold et al. [3].

6.3. *Summary*

It is worth noting that self-avoiding and self-repelling paths offer a host of challenging mathematical problems. Many of them are also of interest in applications, and conversely, the more or less heuristic and often practically successful arguments in the applied sciences — numerical, physical, chemical — provide interesting stimuli and suggestions for further mathematical exploration.

Acknowledgments

This work has financed by Portuguese national funds through FCT - Fundação para a Ciência e Tecnologia within the project PTDC/MAT-STA/1284/2012. LS gratefully acknowledges a travel grant from the Alexander-von-Humboldt Foundation.

References

1. S. Albeverio, M. Roeckner: Dirichlet forms, quantum fields and stochastic quantization. In: Stochastic Analysis, Path Integration and Dynamics, K. D. Elworthy, J. C. Zambrini (Eds.), Wiley, New York, (1989).
2. S. Albeverio, Y.g. Hu, M. Röckner, X. Y. Zhou, Stochastic Quantization of the Two-Dimensional Polymer Measure. *Appl. Math. Optimiz.* **40**, 341-354 (1999).
3. G. Besold, H. Guo, M. J. Zuckermann, Off Lattice Monte Carlo Simulation of the Discrete Edwards Model. *J. Polymer Science* **B 38**, 1053-1068 (2000).
4. F. Biagini, Y. Hu, B. Oksendal, T. Zhang, *Stochastic Calculus for Fractional Brownian Motion and Applications.* Springer, Berlin, 2007.
5. W. Bock, M.-J. Oliveira, J. L. da Silva, L. Streit, Polymer Measure: Varadhan's Renormalization Revisited. *Rev. Math. Phys.* **27**, 1550009 (2015).
6. W. Bock, J. Bornales, C. O. Cabahug, S. Eleutério, L. Streit, Scaling properties of weakly self-avoiding fractional Brownian motion in one dimension. *J. Stat. Phys.* **161**, 1151–1162 (2015).
7. E. Bolthausen, On the construction of the three-dimensional polymer measure. *Probab. Theory Related Fields* **97**, 81–101 (1993).
8. E. Bolthausen, Large Deviations and Interacting Random Walks, Lectures on probability theory and statistics (Saint-Flour, 1999), pp. 1–124, Lecture Notes in Math. 1741, Springer-Verlag, Berlin 2002.
9. J. Bornales, M.-J. Oliveira, L. Streit, Self-repelling fractional Brownian motion - a generalized Edwards model for chain polymers. In L. Accardi, W. Freudenberg, M. Ohya (Eds.), Quantum Probability and White Noise Analysis 30, 389–401. World Scientific Singapore, 2013.
10. J. des Cloizeaux, G. Jannik, Polymers in Solutions: Their Modelling and Structure. Oxford University Press 1990.
11. R. Dekeyser, A. Maritan, A. Stella, Excluded-volume effects in linear polymers: Universality of generalized self-avoiding walks. *Phys. Rev. B* **31**, 4659-4662 (1985).
12. R. Dekeyser, A. Maritan, A. Stella, Random walks with intersections: Static and dynamic fractal properties. *Phys. Rev. A* **36**, 2338-2351 (1987).
13. C. Domb, G. S. Joyce, Cluster expansion for a polymer chain. *Journal of Physics* **C 5**, 956-976 (1972).
14. S. F. Edwards, The statistical mechanics of polymers with excluded volume. *Proc. Roy. Soc.* **85**, 613-624 (1965).
15. M. Faria, T. Hida, L. Streit, H. Watanabe, Intersection Local Times as Generalized White Noise Functional. *Acta Appl. Math.* **46**, 351 (1997).
16. M. E. Fisher, Shape of a self-avoiding walk or polymer chain. *J. Chem. Physics* **44**, 616 (1966).
17. P. J. Flory, *Principles of Polymer Chemistry.* Cornell University Press, 1953
18. P. G. de Gennes, Scaling Concepts in Polymer Physics. Cornell Univ. Press, Ithaca, NY. (1979).
19. M. Grothaus, M. J. Oliveira, J.-L. Silva, L. Streit: Self-avoiding fBm - The Edwards model. *J. Stat. Phys.* **145**, 1513–1523 (2011).

20. R. van der Hofstad, W. König, A Survey of One-Dimensional Random Polymers. *J. Stat. Physics* **103**, 915-944 (2001).

21. W. Koenig, A central limit theorem for a one-dimensional polymer measure, *Ann. Probab.* **24** 1012–1035 (1996).

22. M. K. Kosmas, K. F. Freed, On scaling theories of polymer solutions. *J. Chem. Phys.* **69**, 3647-3659 (1978).

23. S. Kusuoka, *On the path property of Edwards' model for long polymer chains in three dimensions*, in: S. Albeverio (Ed.), Proc. Bielefeld Conf. on Infinite Dimensional Analysis and Stochastic Processes, Pitman Res. Notes MathSci., vol. 124, Wiley, New York (1985), pp. 48–65.

24. N. Madras, G. Slade, *The Self-Avoiding Walk*. Birkhäuser, 1996.

25. Y. Mishura: *Stochastic Calculus for Fractional Brownian Motion and Related Processes*. Springer LNM 1929 (2008).

26. A. Pelissetto, E. Vicari, Critical phenomena and renormalization-group theory. *Phys. Reports* **368**, 549–727 (2002).

27. S. Mendonça, L. Streit, Multiple Intersection Local Times in Terms of White Noise. *IDAQP* **4**, 533 (2001).

28. S. Redner, Distribution functions in the interior of polymer chains. *J. Phys. A: Math. Gen.* **13**, 3525–3541 (1980).

29. M. Rubinstein, R. H. Colby, *Polymer Physics*. Oxford Univ. Press, 2003, 2012.

30. M. Talagrand, Multiple points of trajectories of multiparameter fractional Brownian motion. *Probab. Theory Related Fields* **112**, 545–563 (1998).

31. S. R. S. Varadhan: Appendix to *"Euclidean quantum field theory"* by K. Symanzik, in: R. Jost, ed., *Local Quantum Theory*, Academic Press, New York, p. 285 (1970).

32. J. Westwater, On Edwards' model for polymer chains. *Comm. Math. Phys.* **72**, 131–174 (1980).

33. J. Westwater, On Edwards' model for polymer chains III, Borel summability. *Comm. Math. Phys.* **84**, 459–470 (1982).

NORMAL APPROXIMATION FOR WHITE NOISE FUNCTIONALS BY STEIN'S METHOD AND HIDA CALCULUS

Louis H. Y. Chen

Department of Mathematics
National University of Singapore
10 Lower Kent Ridge Road, Singapore 119076
matchyl@nus.edu.sg

Yuh-Jia Lee

Department of Applied Mathematics
National University of Kaohsiung
Kaohsiung, Taiwan 811
yuhjialee@gmail.com

Hsin-Hung Shih

Department of Applied Mathematics
National University of Kaohsiung
Kaohsiung, Taiwan 811
hhshih@nuk.edu.tw

In this chapter, we establish a framework for normal approximation for white noise functionals by Stein's method and Hida calculus. Our work is inspired by that of Nourdin and Peccati (*Probab. Theory Relat. Fields* 145 (2009), 75-118), who combined Stein's method and Malliavin calculus for normal approximation for functionals of Gaussian processes.

1. Introduction

Stein's method, introduced by C. Stein [39] in his 1972 paper, is a powerful way of determining the accuracy of normal approximation to the distribution of a sum of dependent random variables. It has been extended to approximations by a broad class of other probability distributions such as the Poisson, compound Poisson, and Gamma distributions, and to approximations on finite as well as infinite dimensional spaces. The results of these approximations have been extensively applied in a wide range of other fields such as the theory of random graphs, computational molecular biology, etc. For further details, see [1, 2, 3, 6, 35, 36, 40] and the references cited therein.

Analysis on infinite-dimensional Gaussian spaces has been formulated in terms of Malliavin calculus and Hida calculus. The former, introduced by Malliavin ([29]), studies the calculus of Brownian functionals and their applications on the classical Wiener space. The connection between Stein's method and Malliavin calculus has been explored by Nourdin and Peccati (see [31]). They developed a theory of normal approximation on infinite-dimensional Gaussian spaces. In their connection, the Malliavin derivative D plays an important role. See also [7, 9].

Hida calculus, also known as white noise analysis, is the mathematical theory of white noise initiated by T. Hida in his 1975 Carleton Mathematical Lecture Notes [14]. Let $\{B(t); \, t \in \mathbb{R}\}$ be a standard Brownian motion and let the white noise $\dot{B}(t) \equiv dB(t)/dt, t \in \mathbb{R}$, be represented by generalized functions. By regarding the collection $\{\dot{B}(t); \, t \in \mathbb{R}\}$ as a coordinate system, Hida defined and studied generalized white noise functionals $\varphi(\dot{B}(t), \, t \in \mathbb{R})$ through their U-functionals. We refer the interested reader to [15, 16, 19, 38].

The objective of this chapter is to develop a connection between Stein's method and Hida calculus for normal approximation for white noise functionals (see Section 5). Our approach is analogous to that for the connection between Stein's method and Malliavin calculus as established by Nourdin and Peccati [31]. The connection between Stein's method and Hida calculus will be built on the expression of the number operator (or the Ornstein-Ulenbeck operator) in terms of the Hida derivatives through integration by parts techniques (see Section 4). The difficulty that we have encountered so far is that the Hida derivative ∂_t, that is the $\dot{B}(t)$-differentiation, cannot be defined on all square-integrable white noise functionals in (L^2). Extending the domain of ∂_t to a larger subclass of (L^2) and studying the regularity of ∂_t will be a key contribution in our chapter.

At the time of completing this chapter we came to know about the PhD thesis of Chu [11], in which he developed normal approximation (in Wasserstein distance) for Lévy functionals by applying Stein's method and Hida calculus. He achieved this by using the white noise approach of Lee and Shih [28].

We list some notations which will be often used in this chapter.

Notation 1.1:

(1) For a real locally convex space V, $\mathscr{C}V$ denotes its complexification. If $(V, |\cdot|_V)$ is a real Hilbert space, then $\mathscr{C}V$ is a complex Hilbert space with the $|\cdot|_{\mathscr{C}V}$-norm given by $|\phi|^2_{\mathscr{C}V} = |\phi_1|^2_V + |\phi_2|^2_V$ for any $\phi = \phi_1 + i\phi_2$, $\phi_1, \phi_2 \in V$. Specially, for $V = S_p$ with $|\cdot|_p$-norm (see Section 3 for the definition), we will still use $|\cdot|_p$ to denote $|\cdot|_{\mathscr{C}S_p}$.

(2) The symbol $(\cdot, \cdot)$ denotes the S'-S, $\mathscr{C}S'$-$\mathscr{C}S$, S_{-p}-S_p, or $\mathscr{C}S_{-p}$-$\mathscr{C}S_p$ pairing.

(3) For a n-linear operator T on $X \times \cdots \times X$, Tx^n means $T(x, \ldots, x)$ as well as $Tx^{n-1}y = T(x, \ldots, x, y)$, $x, y \in X$, where X is a real or complex locally convex space.

(4) The constant ω_r with $r > 0$ is given by the square root of $\sup\{n/2^{2nr}; n \in \mathbb{N}_0\}$, where $\mathbb{N}_0 = \mathbb{N} \cup \{0\}$.

2. Stein's Method

In this section we give a brief exposition of the basics of Stein's method for normal approximation.

2.1. *From characterization to approximation*

In his 1986 monograph [39], Stein proved the following characterization of the normal distribution.

Proposition 2.1: *(Stein's lemma) The following are equivalent.*
(i) $W \sim \mathcal{N}(0, 1)$;
(ii) $\mathbb{E}[f'(W) - Wf(W)] = 0$ for all $f \in \mathcal{C}^1_B$.

Proof: By integration by parts, (i) implies (ii). If (ii) holds, solve

$$f'(w) - wf(w) = h(w) - \mathbb{E}h(Z)$$

where $h \in C_B$ and Z has the standard normal distribution (denoted by $Z \sim N(0, 1)$). Its bounded unique solution f_h is given by

$$f_h(w) = -e^{\frac{1}{2}w^2} \int_w^\infty e^{-\frac{1}{2}t^2}[h(t) - \mathbb{E}h(Z)]dt$$

$$= e^{\frac{1}{2}w^2} \int_{-\infty}^w e^{-\frac{1}{2}t^2}[h(t) - \mathbb{E}h(Z)]dt. \tag{2.1}$$

Using $\int_w^\infty e^{-\frac{1}{2}t^2} dt \leq w^{-1}e^{-\frac{1}{2}w^2}$ for $w > 0$, we can show that $f_h \in \mathcal{C}^1_B$ with $\|f_h\|_\infty \leq \sqrt{2\pi e}\|h\|_\infty$ and $\|f'_h\|_\infty \leq 4\|h\|_\infty$. Substituting f_h for f in (ii) leads to

$$\mathbb{E}h(W) = \mathbb{E}h(Z) \quad \text{for} \quad h \in \mathcal{C}_B.$$

This proves (i). □

The equation

$$\mathbb{E}Wf(W) = \mathbb{E}f'(W) \tag{2.2}$$

for all $f \in \mathcal{C}_B^1$, which characterizes the standard normal distribution, is called the Stein identity for normal distribution. In fact, if $W \sim N(0,1)$, (2.2) holds for absolutely continuous f such that $\mathbb{E}|f'(W)| < \infty$.

Let W be a random variable with $\mathbb{E}W = 0$ and $\text{Var}(W) = 1$. Proposition 2.1 suggests that the distribution of W is "close" to $N(0,1)$ if and only if

$$\mathbb{E}[f'(W) - Wf(W)] \simeq 0$$

for $f \in \mathcal{C}_B^1$. How "close" the distribution of W is to the standard normal distribution may be quantified by determining how close $\mathbb{E}[f'(W) - Wf(W)]$ is to 0. To this end we define a distance between the distribution of W and the standard normal distribution as follows.

$$d_{\mathcal{G}}(W, Z) := \sup_{h \in \mathcal{G}} |\mathbb{E}h(W) - \mathbb{E}h(Z)| \tag{2.3}$$

where $\mathcal{G}$ is a separating class and the distance $d_{\mathcal{G}}$ is said to be induced by $\mathcal{G}$. By a separating class $\mathcal{G}$, we mean a class of Borel measurable real-valued functions defined on $\mathbb{R}$ such that two random variables, X and Y, have the same distribution if $\mathbb{E}h(X) = \mathbb{E}h(Y)$ for $h \in \mathcal{G}$. Such a separating class contains functions h for which both $\mathbb{E}h(X)$ and $\mathbb{E}h(Y)$ exist.

Let f_h be the solution, given by (2.1), of the Stein's equation

$$f'(w) - wf(w) = h(w) - \mathbb{E}h(Z) \tag{2.4}$$

where $h \in \mathcal{G}$. Then we have

$$d_{\mathcal{G}}(W, Z) = \sup_{h \in \mathcal{G}} |\mathbb{E}[f_h'(W) - Wf_h(W)]|. \tag{2.5}$$

So bounding the distance $d_{\mathcal{G}}(W, Z)$ is equivalent to bounding $\sup_{h \in \mathcal{G}} |\mathbb{E}[f_h'(W) - Wf_h(W)]|$, for which we need to study the boundedness properties of f_h and the probabilistic structure of W.

The following three separating classes of Borel measurable real-valued functions defined on $\mathbb{R}$ are of interest in normal approximation.

$$\mathcal{G}_W := \{h; |h(u) - h(v)| \leq |u - v|\},$$

$$\mathcal{G}_K := \{h; h(w) = 1 \text{ for } w \leq x \text{ and } = 0 \text{ for } w > x, x \in \mathbb{R}\},$$

$$\mathcal{G}_{TV} := \{h; h(w) = I(w \in A), A \text{ is a Borel subset of } \mathbb{R}\}.$$

The distances induced by these three separating classes are respectively called the Wasserstein distance, the Kolmogorov distance, and the total variation distance. It is customary to denote $d_{\mathscr{G}_W}$, $d_{\mathscr{G}_K}$ and $d_{\mathscr{G}_{TV}}$ respectively by d_W, d_K and d_{TV}.

Since for each h such that $|h(x) - h(y)| \leq |x - y|$, there exists a sequence of $h_n \in \mathcal{C}^1$ with $\|h'_n\|_\infty \leq 1$ such that $\|h_n - h\|_\infty \to 0$ as $n \to \infty$, we have

$$d_W(W, Z) = \sup_{h \in \mathcal{C}^1, \|h'\|_\infty \leq 1} |\mathbb{E}h(W) - \mathbb{E}h(Z)|. \tag{2.6}$$

By an application of Lusin's theorem, we also have

$$d_{TV}(W, Z) = \sup_{h \in \mathcal{C}, 0 \leq h \leq 1} |\mathbb{E}h(W) - \mathbb{E}h(Z)|. \tag{2.7}$$

It is also known that

$$d_{TV}(W, Z) = \frac{1}{2} \sup_{\|h\|_\infty \leq 1} |\mathbb{E}h(W) - \mathbb{E}h(Z)|$$

$$= \frac{1}{2} \sup_{h \in \mathcal{C}, \|h\|_\infty \leq 1} |\mathbb{E}h(W) - \mathbb{E}h(Z)|.$$

It is generally much harder to obtain an optimal bound on the Kolmogorov distance than on the Wasserstein distance. There is a discussion on this and examples of bounding the Wasserstein distance and the Kolmogorov distance are given in Chen [7].

We now state a proposition that concerns the boundedness properties of the solution f_h, given by (2.1), of the Stein equation (2.4) for h either bounded or absolutely continuous with bounded h'. The use of these boundedness properties is crucial for bounding the Wasserstein, Kolmogorov and total variation distances.

Proposition 2.2: *Let f_h be the unique solution, given by (2.1), of the Stein equation (2.4), where h is either bounded or absolutely continuous.*

1. If h is bounded, then

$$\|f_h\|_\infty \leq \sqrt{\pi/2} \|h - \mathbb{E}h(Z)\|_\infty, \quad \|f'_h\|_\infty \leq 2\|h - \mathbb{E}h(Z)\|_\infty. \tag{2.8}$$

2. If h is absolutely continuous with bounded h', then

$$\|f_h\|_\infty \leq 2\|h'\|_\infty, \quad \|f'_h\|_\infty \leq \sqrt{2/\pi}\|h'\|_\infty, \quad \|f''_h\|_\infty \leq 2\|h'\|_\infty. \tag{2.9}$$

3. If $h = I_{(-\infty, x]}$ where $x \in \mathbb{R}$, then, writing f_h as f_x,

$$0 < f_x(w) \leq \sqrt{2\pi}/4, \quad |w f_x(w)| \leq 1, \quad |f'_x(w)| \leq 1, \tag{2.10}$$

and for all $w, u, v \in \mathbb{R}$,

$$|f'_x(w) - f'_x(v)| \leq 1, \qquad (2.11)$$

$$|(w+u)f_x(w+u) - (w+v)f_x(w+v)| \leq (|w| + \sqrt{2\pi}/4)(|u| + |v|). \quad (2.12)$$

4. If $h = h_{x,\epsilon}$ *where* $x \in \mathbb{R}$, $\epsilon > 0$, *and*

$$h_{x,\epsilon}(w) = \begin{cases} 1, & w \leq x, \\ 0, & w \geq x + \epsilon, \\ 1 + \epsilon^{-1}(x - w), & x < w < x + \epsilon, \end{cases}$$

then, writing f_h *as* $f_{x,\epsilon}$, *we have for all* $w, v, t \in \mathbb{R}$,

$$0 \leq f_{x,\epsilon}(w) \leq 1, \quad |f'_{x,\epsilon}(w)| \leq 1, \quad |f'_{x,\epsilon}(w) - f'_{x,\epsilon}(v)| \leq 1 \qquad (2.13)$$

and

$$|f'_{x,\epsilon}(w+t) - f'_{x,\epsilon}(w)|$$
$$\leq (|w| + 1)|t| + \frac{1}{\epsilon} \int_{t \wedge 0}^{t \vee 0} I(x \leq w + u \leq x + \epsilon) du$$
$$\leq (|w| + 1)|t| + I(x - 0 \vee t \leq w \leq x - 0 \wedge t + \epsilon). \qquad (2.14)$$

Except for (2.14), which can be found on page 2010 in Chen and Shao [10], the bounds in Proposition 2.2 and their proofs can be found in Lemmas 2.3, 2.4 and 2.5 in Chen, Goldstein and Shao [8].

2.2. *Stein identities and error terms*

Let W be a random variable with $\mathbb{E}W = 0$ and $\mathrm{Var}(W) = 1$. In addition to using the boundedness properties of the solution f_h of the Stein equation (2.4), we also need to exploit the probabilistic structure of W, in order to bound the error term in (2.5). This is done through the construction of a Stein identity for W for normal approximation. This is perhaps best understood by looking at a specific example.

Let $X_1, \ldots, X_n$ be independent random variables with $\mathbb{E}X_i = 0$, $\mathrm{Var}(X_i) = \sigma_i^2$ and $\mathbb{E}|X_i|^3 < \infty$. Let $W = \sum_{i=1}^{n} X_i$ and $W^{(i)} = W - X_i$ for $i = 1, \ldots, n$. Assume that $\mathrm{Var}(W) = 1$, which implies $\sum_{i=1}^{n} \sigma_i^2 = 1$. Let $f \in C_B^2$. Using the independence among the X_i and the property that

$\mathbb{E}X_i = 0$, we have

$$\mathbb{E}Wf(W) = \sum_{i=1}^{n} \mathbb{E}X_i f(W) = \sum_{i=1}^{n} \mathbb{E}X_i[f(W^{(i)} + X_i) - f(W^{(i)})]$$

$$= \sum_{i=1}^{n} \mathbb{E} \int_0^{X_i} X_i f'(W^{(i)} + t)dt$$

$$= \sum_{i=i}^{n} \mathbb{E} \int_{-\infty}^{\infty} f'(W^{(i)} + t)\widehat{K}_i(t)dt$$

$$= \sum_{i=i}^{n} \mathbb{E} \int_{-\infty}^{\infty} f'(W^{(i)} + t)K_i(t)dt \qquad (2.15)$$

where

$$\widehat{K}_i(t) = X_i[I(X_i > t > 0) - I(X_i < t < 0)],$$
$$K_i(t) = \mathbb{E}\widehat{K}_i(t).$$

It can be shown that for each i, $\sigma_i^{-2}K_i$ is a probability density function. So (2.15) can be rewritten as

$$\mathbb{E}Wf(W) = \sum_{i=1}^{n} \sigma_i^2 \mathbb{E}f'(W^{(i)} + T_i) \qquad (2.16)$$

where $T_1, \ldots, T_n, X_1, \ldots, X_n$ are independent and T_i has the density $\sigma_i^{-2}K_i$. Both the equations (2.15) and (2.16) are Stein identities for W for normal approximation. From (2.16) we obtain

$$\mathbb{E}[f'(W) - Wf(W)] = \sum_{i=1}^{n} \sigma_i^2 \mathbb{E}[f'(W) - f'(W^{(i)})]$$

$$- \sum_{i=1}^{n} \sigma_i^2 \mathbb{E}[f'(W^{(i)} + T_i) - f'(W^{(i)})] \quad (2.17)$$

where the error terms on the right hand side provide an expression for the deviation of $\mathbb{E}[f'(W) - Wf(W)]$ from 0. Now let f be f_h where $h \in \mathcal{C}^1$ such that $\|h'\|_\infty \le 1$. Applying Taylor expansion to (2.17), and using (2.6) and

(2.9), we obtain

$$
d_W(W, Z) \leq \sum_{i=1}^{n} \sup_{h \in \mathcal{C}^1, \|h'\|_\infty \leq 1} \sigma_i^2 \mathbb{E}|X_i| \, \|f_h''\|_\infty
$$

$$
+ \sum_{i=1}^{n} \sup_{h \in \mathcal{C}^1, \|h'\|_\infty \leq 1} \sigma_i^2 \mathbb{E}|T_i| \, \|f_h''\|_\infty
$$

$$
\leq 2 \sum_{i=1}^{n} \sigma_i^2 \mathbb{E}|X_i| + 2 \sum_{i=1}^{n} \sigma_i^2 \mathbb{E}|T_i|.
$$

Since $\sigma_i^2 \mathbb{E}|X_i| \leq (\mathbb{E}|X_i|^3)^{2/3})(\mathbb{E}|X_i|^3)^{1/3} = \mathbb{E}|X_i|^3$ and $\sigma_i^2 \mathbb{E}|T_i| = \sigma_i^2(1/2\sigma_i^2)\mathbb{E}|X_i|^3 = (1/2)\mathbb{E}|X_i|^3$, we have

$$
d_W(W, Z) \leq 3 \sum_{i=1}^{n} \mathbb{E}|X_i|^3. \tag{2.18}
$$

The bound in (2.18) is of optimal order. However, it is much harder to obtain a bound of optimal order for the Kolmogorov distance. Proofs of such a bound on $d_K(W, Z)$ can be found in Chen [7] and Chen, Goldstein and Shao [8]. Stein's identities for locally dependent random variables and other dependent random variables can be found in Chen and Shao [10] and Chen, Goldstein and Shao [8].

2.3. *Integration by parts*

Let W be a random variable with $\mathbb{E}W = 0$ and $\mathrm{Var}(W) = 1$. In many situations of normal approximation, the Stein identity for W takes the form

$$
\mathbb{E}Wf(W) = \mathbb{E}Tf'(W) \tag{2.19}
$$

where T is random variable defined on the same probability as W such that $\mathbb{E}|T| < \infty$, and f an absolutely continuous function for which the expectations exist. Typical examples of such situations are cases when W is a functional of Gaussian random variables or of a Gaussian process. In such situations, the Stein identity (2.19) is often constructed using integration by parts as in the case of the Stein identity for $N(0, 1)$.

By letting $f(w) = w$, we obtain $\mathbb{E}T = \mathbb{E}W^2 = 1$. Let $\mathscr{F}$ be a σ-algebra with respect to which W is measurable. From (2.19),

$$
\mathbb{E}[f'(W) - Wf(W)] = \mathbb{E}[f'(W)(1 - T)]
$$

$$
= \mathbb{E}[f'(W)\mathbb{E}(1 - T|\mathscr{F})].
$$

Now let $f = f_h$ where $h \in \mathscr{G}$, a separating class of functions. Assume that $\|f_h'\|_\infty < \infty$. Then by (2.3),

$$
\begin{aligned}
d_{\mathscr{G}}(W, Z) &= \sup_{h \in \mathscr{G}} |\mathbb{E}[f_h'(W) - W f_h(W)]| \\
&\le \left(\sup_{h \in \mathscr{G}} \|f_h'\|_\infty \right) \mathbb{E}|\mathbb{E}(1 - T|\mathscr{F})| \\
&\le \left(\sup_{h \in \mathscr{G}} \|f_h'\|_\infty \right) \sqrt{\mathrm{Var}(\mathbb{E}(T|\mathscr{F}))},
\end{aligned}
$$

where for the last inequality it is assumed that $\mathbb{E}(T|\mathscr{F})$ is square integrable. By Proposition 2.2,

$$
\sup_{h \in \mathscr{G}} \|f_h'\|_\infty \le
\begin{cases}
\sqrt{2/\pi}, & \mathscr{G} = \mathscr{G}_W, \\
1, & \mathscr{G} = \mathscr{G}_K, \\
2, & \mathscr{G} = \mathscr{G}_{TV}.
\end{cases}
\tag{2.20}
$$

This implies

$$
d(W, Z) \le \theta \mathbb{E}|\mathbb{E}(1 - T|\mathscr{F})| \le \theta \sqrt{\mathrm{Var}(\mathbb{E}(T|\mathscr{F}))}
\tag{2.21}
$$

where

$$
\theta =
\begin{cases}
\sqrt{2/\pi}, & d(W, Z) = d_W(W, Z), \\
1, & d(W, Z) = d_K(W, Z), \\
2, & d(W, Z) = d_{TV}(W, Z).
\end{cases}
\tag{2.22}
$$

For the rest of this section, we will present two approaches to the construction of the Stein identity (2.19).

Let $Z \sim N(0, 1)$ and let ψ be an absolutely continuous function such that $\mathbb{E}\psi(Z) = 0$, $\mathrm{Var}(\psi(Z)) = 1$ and $\mathbb{E}\psi'(Z)^2 < \infty$. Define $W = \psi(Z)$. Following Chatterjee [5], we use Gaussian interpolation to construct a Stein identity for $\psi(Z)$. Let f be absolutely continuous with bounded derivative, which implies $|f(w)| \le C(1 + |w|)$ for some $C > 0$. Let Z' be an independent copy of Z and let $W_t = \psi(\sqrt{t}Z + \sqrt{1 - t}Z')$ for $0 \le t \le 1$. Then we have

$$
\begin{aligned}
\mathbb{E}W f(W) &= \mathbb{E}(W_1 - W_0) f(W) = \mathbb{E} \int_0^1 f(W) \frac{\partial W_t}{\partial t} dt \\
&= \mathbb{E} \int_0^1 f(W) \left(\frac{Z}{2\sqrt{t}} - \frac{Z'}{2\sqrt{1 - t}} \right) \psi'(\sqrt{t}Z + \sqrt{1 - t}Z') dt. \quad (2.23)
\end{aligned}
$$

Let

$$
\begin{aligned}
U_t &= \sqrt{t}Z + \sqrt{1 - t}Z', \\
V_t &= \sqrt{1 - t}Z - \sqrt{t}Z'.
\end{aligned}
$$

Then $U_t \sim N(0,1)$, $V_t \sim N(0,1)$, and U_t and V_t are independent. This together with $|f(w)| \leq C(1 + |w|)$ implies the integrability of the right hand side of (2.23). Solving for Z, we obtain

$$Z = \sqrt{t}U_t + \sqrt{1-t}V_t.$$

Equation (2.23) can be rewritten as

$$\mathbb{E}Wf(W) = \frac{1}{2} \int_0^1 \frac{1}{\sqrt{t(1-t)}} \mathbb{E}f(\psi(\sqrt{t}U_t + \sqrt{1-t}V_t))V_t\psi'(U_t)$$

$$= \frac{1}{2} \int_0^1 \frac{1}{\sqrt{t}} \mathbb{E}f'(W)\psi'(Z)\psi'(U_t)dt$$

$$= \frac{1}{2}\mathbb{E}\left[f'(W)\mathbb{E}\left(\int_0^1 \frac{1}{\sqrt{t}}\psi'(Z)\psi'(\sqrt{t}Z + \sqrt{1-t}Z')dt \Big| Z \right) \right]$$

where for the second equality we used the indepedence of U_t and V_t, and applied the characterization equation for $N(0,1)$ to V_t. Note that the characterizaton equation is obtained by integration by parts. Hence we have for absolutely continuous f with bounded derivative,

$$\mathbb{E}Wf(W) = \mathbb{E}T(Z)f'(W) \qquad (2.24)$$

where

$$T(x) = \int_0^1 \frac{1}{2\sqrt{t}}\mathbb{E}\left[\psi'(x)\psi'(\sqrt{t}x + \sqrt{1-t}Z') \right] dt. \qquad (2.25)$$

In [5], Chatterjee obtained a multivariate version of (2.24) where $\psi : \mathbb{R}^d \longrightarrow \mathbb{R}$ for $d \geq 1$.

Here is a simple application of (2.21) and (2.24). Let $X_1, \ldots, X_n$ be independent and identically distributed as $N(0,1)$. Let

$$W = \frac{\sum_{i=1}^n (X_i^2 - 1)}{\sqrt{2n}}.$$

The random variable W has the standardized χ^2 distribution with n degrees of freedom. Let $\psi(X_i) = \frac{X_i^2 - 1}{\sqrt{2n}}$. Then $W = \sum_{i=1}^n \psi(X_i)$. Let $W^{(i)} = W - \psi(X_i)$ and let $X_1', \ldots, X_n'$ be an independent copy of $X_1, \ldots, X_n$. By the independence of $X_1, \ldots, X_n$, and by (2.24) and (2.25),

$$\mathbb{E}Wf(W) = \sum_{i=1}^n \mathbb{E}\psi(X_i)f'(W^{(i)} + g(X_i)) = \sum_{i=1}^n \mathbb{E}T(X_i)f'(W)$$

$$= \mathbb{E}Tf'(W).$$

where

$$T = \sum_{i=1}^{n} T(X_i),$$

$$T(X_i) = \int_0^1 \frac{1}{2\sqrt{t}} \mathbb{E}\left(\psi'(X_i)\psi'(\sqrt{t}X_i + \sqrt{1-t}X_i')\big|X_i\right) dt$$

$$= \int_0^1 \frac{1}{n\sqrt{t}} \mathbb{E}\left(X_i(\sqrt{t}X_i + \sqrt{1-t}X_i')\big|X_i\right) dt = \frac{X_i^2}{n}.$$

Therefore

$$\mathrm{Var}(T) = \sum_{i=1}^{n} \mathrm{Var}\left(\frac{X_i^2}{n}\right) = \frac{2}{n}.$$

By (2.21) and since $\mathrm{Var}(\mathbb{E}(T|\mathscr{F})) \le \mathrm{Var}(T)$, it follows that

$$d_W(W, Z) \le \frac{2}{\sqrt{\pi}} \frac{1}{\sqrt{n}}, \quad d_K(W, Z) \le \frac{\sqrt{2}}{\sqrt{n}}, \quad d_{TV}(W, Z) \le \frac{2\sqrt{2}}{\sqrt{n}}. \quad (2.26)$$

Since $d_K(W, Z) \le d_{TV}(W, Z)$, this result is stronger and more general than what can be deduced from the Berry-Esseen theorem, which yields only the Kolmogorov bound.

We now present another approach to the construction of the Stein identity (2.24). Consider $\mathcal{L}^2 = L^2\left(\mathbb{R}, \frac{1}{\sqrt{2\pi}} e^{-\frac{x^2}{2}} dx\right)$ endowed with the inner product,

$$\langle f, g \rangle = \int_{\mathbb{R}} f(x)g(x) \frac{1}{\sqrt{2\pi}} e^{-\frac{x^2}{2}} dx.$$

Let L be the Ornstein-Uhlenbeck operator defined on $\mathcal{D}^2 \subset \mathcal{L}^2$, where

$$\mathcal{D}^2 = \{g : g' \text{ is absolutely continuous, } g, g', g'' \in \mathcal{L}^2\}.$$

That is,

$$L = \frac{d^2}{d^2 x} - x \frac{d}{dx}.$$

Let $f \in \mathcal{L}^2$ be absolutely continuous such that $f' \in \mathcal{L}^2$ and let $g \in \mathrm{Dom}(L)$. Assume that $g'(x)f(x)e^{-x^2/2} \to 0$ as $x \to \pm$. Then we have the integration by parts formula,

$$\langle Lg, f \rangle = -\langle g', f' \rangle. \quad (2.27)$$

Let $\psi \in \mathcal{L}^2$ be absolutely continuous such that $\mathbb{E}\psi(Z) = 0$, $\mathrm{Var}(\psi(Z)) = 1$ and ψ' is bounded by a polynomial. A solution g_ψ of the equation

$$Lg = \psi$$

is given by

$$g_\psi(x) = L^{-1}\psi = -\int_0^\infty P_t\psi dt$$

$$= -\int_0^\infty \mathbb{E}\psi(e^{-t}x + \sqrt{1 - e^{-2t}}Z)dt$$

where L^{-1} is a pseudo-inverse of L, $(P_t)_{t\geq 0}$ is the Ornstein-Uhlenbeck se-migroup defined on $\mathcal{L}^2$, and Z and Z' are independent, each distributed as $N(0,1)$. By the integration by parts formula (2.27), we have for absolutely continuous f such that $\|f'\|_\infty < \infty$,

$$\mathbb{E}\psi(Z)f(\psi(Z))$$
$$= \langle \psi, f(\psi) \rangle = \langle LL^{-1}\psi, f(\psi) \rangle$$
$$= -\langle (L^{-1}\psi)', f'(\psi)\psi' \rangle = -\langle g'_\psi, f'(\psi)\psi' \rangle$$
$$= \mathbb{E}\int_0^\infty e^{-s}f'(\psi(Z))\psi'(Z)\mathbb{E}\left(\psi'(e^{-s}x + \sqrt{1 - e^{-2s}}Z')\big|Z\right)ds$$
$$= \mathbb{E}\int_0^1 \frac{1}{2\sqrt{t}}f'(\psi(Z))\psi'(Z)\mathbb{E}\left(\psi'(\sqrt{t}Z + \sqrt{1 - t}Z')\big|Z\right)dt \quad (2.28)$$

where the last equality follows from the change of variable, $t = e^{-2s}$. By using the fact that polynomials are dense in $\mathcal{L}^2$, (2.28) can be shown to hold for $\mathbb{E}\psi'(Z)^2 < \infty$. By letting $W = \psi(Z)$, the identity (2.28) is indeed the same as (2.24).

This approach of using the integration by parts formula (2.27) to construct a Stein identity is a special case of that of Nourdin and Peccati [31], who considered $L^2(\Omega) = L^2(\Omega, \mathscr{F}, P)$, where $\mathscr{F}$ is the complete σ-algebra generated by a standard Brownian motion defined on Ω. The integration by parts formula in this setting is

$$\langle LF, G \rangle_{L^2(\Omega)} = -\mathbb{E}[\langle DF, DG \rangle_{L^2(\mathbb{R}_+, dx)}] \quad (2.29)$$

where $F \in \mathbb{D}^{2,2}$, $G \in \mathbb{D}^{1,2}$, L is the Ornstein-Uhlenbeck operator defined on $\mathbb{D}^{2,2} \subset L^2(\Omega)$ and D the Malliavin derivative with the domain $\mathbb{D}^{1,2}$.

In this chapter, we will use a similar integration by parts formula in white noise analysis involving the Hida derivative to obtain a general error bound in the normal approximation for white noise functionals. In the process, it is found necessary to extend the domain of the Hida derivative to allow application of the integration by parts formula to the normal approximation.

3. Hida Distributions

In this and the next sections, we will give a brief description of Hida's white noise calculus based on Lee's reformulation on the abstract Wiener space $(\mathcal{S}_0, \mathcal{S}_{-p})$ for $p > \frac{1}{2}$. For more details, see [21, 22, 23, 24].

3.1. *White noise space*

Let $\mathbf{A} = -(d/dt)^2 + 1 + t^2$ be a densely defined self-adjoint operator on the L^2-space $L^2(\mathbb{R}, dt)$ with respect to the Lebesgue measure dt, and $\{h_n;\ n \in \mathbb{N}_0 \equiv \mathbb{N} \cup \{0\})\}$ be a complete orthonormal set (CONS for abbreviation) for $L^2(\mathbb{R}, dt)$, consisting of all Hermite functions on $\mathbb{R}$, formed by the eigenfunctions of $\mathbf{A}$ with corresponding eigenvalues $2n + 2$, $n \in \mathbb{N}_0$, where

$$h_n(t) = \frac{1}{\sqrt{\sqrt{\pi} 2^n n!}} H_n(t)\, e^{-t^2/2},$$

$H_n(t) = (-1)^n e^{t^2} \frac{d^n}{dt^n} e^{-t^2}$ being the Hermite polynomial of the degree n.

Let $\mathcal{S}$ be the Schwartz space of real-valued, rapidly decreasing, and infinitely differentiable functions on $\mathbb{R}$ with its dual $\mathcal{S}'$, the spaces of tempered distributions. For each $p \in \mathbb{R}$, let $\mathcal{S}_p$ denote the space of all functions f in $\mathcal{S}'$ satisfying the condition that

$$|f|_p^2 \equiv \sum_{n=0}^{\infty} (2n + 2)^{2p} |(f,\, h_n)|^2 < +\infty,$$

where $(\cdot, \cdot)$ always denotes the $\mathcal{S}'$-$\mathcal{S}$ pairing from now on. Then $\mathcal{S}_p$, $p \in \mathbb{R}$, forms a real Hilbert space with the inner product $\langle \cdot, \cdot \rangle_p$ induced by $|\cdot|_p$. The dual space $\mathcal{S}_p'$, $p \in \mathbb{R}$, is unitarily equivalent to $\mathcal{S}_{-p}$. Applying the Riesz representation theorem, we have the continuous inclusions:

$$\mathcal{S} \subset \mathcal{S}_q \subset \mathcal{S}_p \subset L^2(\mathbb{R}, dt) = \mathcal{S}_0 \subset \mathcal{S}_{-p} \subset \mathcal{S}_{-q} \subset \mathcal{S}',$$

where $0 < p < q < +\infty$, $\mathcal{S}$ is the projective limit of $\{\mathcal{S}_p;\ p > 0\}$. In fact, $\mathcal{S}$ is a nuclear space, and thus $\mathcal{S}'$ is the inductive limit of $\{\mathcal{S}_{-p};\ p > 0\}$.

A well known fact is that the Minlos theorem (see [12]) guarantees the existence of the white noise measure μ on $(\mathcal{S}', \mathscr{B}(\mathcal{S}'))$, $\mathscr{B}(\mathcal{S}')$ being the Borel σ-field of $\mathcal{S}'$, the characteristic functional of which is given by

$$\int_{\mathcal{S}'} e^{\mathrm{i}\,(x,\, \eta)}\, \mu(dx) = e^{-\frac{1}{2}|\eta|_0^2}, \quad \text{for all } \eta \in \mathcal{S}, \tag{3.1}$$

where $\mathrm{i} = \sqrt{-1}$. One can easily show that the measurable support of μ is contained in $\mathcal{S}_{-p}$ and μ coincides with the Wiener measure on the abstract Wiener space $(\mathcal{S}_0,\ \mathcal{S}_{-p})$ for $p > \frac{1}{2}$.

As a random variable on $(\mathcal{S}', \mathcal{B}(\mathcal{S}'), \mu)$, $(\cdot, \eta)$ has the normal distribution with mean 0 and variance $|\eta|_0^2$ for any $\eta \in \mathcal{S}$. For each $\rho \in \mathcal{S}_0$, choose a sequence $\{\eta_n\} \subset \mathcal{S}$ so that $\eta_n \to \rho$ in $\mathcal{S}_0$. Then it follows from (3.1) that $\{(\cdot, \eta_n)\}$ forms a Cauchy sequence in $(L^2) \equiv L^2(\mathcal{S}', \mu)$, the L^2-space of all complex-valued square-integrable functionals on $\mathcal{S}'$ with respect to μ. Denote by $\langle \cdot, \rho \rangle$ the L^2-limit of $\{(\cdot, \eta_n)\}$. Then $\langle \cdot, \rho \rangle \sim \mathcal{N}(0, |\rho|_0^2)$ for $\rho \in \mathcal{S}_0$. Consequently, the Brownian motion $B = \{B(t); t \in \mathbb{R}\}$ on $(\mathcal{S}', \mathcal{B}(\mathcal{S}'), \mu)$ can be represented by

$$B(t\,;\,x) = \begin{cases} \langle x, 1_{[0,\,t]} \rangle, & \text{if } t \geq 0 \\ -\langle x, 1_{[t,\,0]} \rangle, & \text{if } t < 0, \ x \in \mathcal{S}'. \end{cases}$$

Taking the time derivative formally, we get $\dot{B}(t\,;\,x) = x(t)$, $x \in \mathcal{S}'$. Thus an element $x \in \mathcal{S}'$ is viewed as a sample path of white noise $\dot{B}(t, x)$ and the space $(\mathcal{S}', \mu)$ is referred to as a white noise space.

3.2. The S-transform

The S-transform $S\varphi$ of $\varphi \in (L^2)$ is a Bargmann-Segal analytic functional on $\mathscr{C}\mathcal{S}_0$ given by

$$S\varphi(\eta) = e^{-\frac{1}{2}\int_{-\infty}^{\infty} \eta(t)^2\, dt} \int_{\mathcal{S}'} \varphi(x)\, e^{\langle x,\, \eta \rangle}\, \mu(dx), \quad \eta \in \mathscr{C}\mathcal{S}_0.$$

We should note that $S\varphi(\eta) = \mu\varphi(\eta)$ for $\eta \in \mathcal{S}_0$, where $\mu\varphi = \mu * \varphi$, the convolution of μ and φ.

Let $\varphi \in (L^2)$ be given. Then it follows from [21, 22] that $D^n S\varphi(0)$ is a symmetric n-linear operator of Hilbert-Schmidt type on $\mathscr{C}\mathcal{S}_0 \times \cdots \times \mathscr{C}\mathcal{S}_0$, where D is the Fréchet derivative of $S\varphi$. Also, it admits the Wiener-Itô decomposition

$$\varphi(x) = \sum_{n=0}^{\infty} \frac{1}{n!} : D^n S\varphi(0)x^n :,$$

and

$$\|\varphi\|_{2,0}^2 \equiv \int_{\mathcal{S}'} |\varphi(x)|^2\, \mu(dx) = \sum_{n=0}^{\infty} \frac{1}{n!} \|D^n S\varphi(0)\|_{\mathcal{HS}^n(\mathscr{C}\mathcal{S}_0)}^2,$$

where $\|\cdot\|_{\mathcal{HS}^n(K)}$ denotes the Hilbert-Schmidt operator norm of a n-linear functional on a Hilbert space K, and for any symmetric n-linear Hilbert-

Schmidt operator T on $\mathscr{C}\mathcal{S}_0 \times \cdots \times \mathscr{C}\mathcal{S}_0$,

$$: Tx^n : \equiv \int_{\mathcal{S}'} T(x + \mathrm{i}\,y)^n \, \mu(dy)$$

$$= \sum_{j_1,\ldots,j_n=0}^{\infty} T(h_{j_1}, \ldots, h_{j_n}) \int_{\mathcal{S}'} \prod_{k=1}^{n} (x + \mathrm{i}\,y, h_{j_k}) \, \mu(dy),$$

which is in (L^2) with $\| : Tx^n : \|_{2,0} = \sqrt{n!} \, \|T\|_{\mathcal{HS}^n(\mathscr{C}\mathcal{S}_0)}$ (see [21]). In fact, the S-transform is a unitary operator from (L^2) onto the Bargmann-Segal-Dwyer space $\mathcal{F}^1(\mathscr{C}\mathcal{S}_0)$ over $\mathscr{C}\mathcal{S}_0$ (see [24]).

Remark 3.1: Let H be a complex Hilbert space. For $r > 0$, denote by $\mathcal{F}^r(H)$ the class of analytic functionals on H with norm $\| \cdot \|_{\mathcal{F}^r(H)}$ such that

$$\|f\|^2_{\mathcal{F}^r(H)} = \sum_{n=0}^{\infty} \frac{r^n}{n!} \|D^n f(0)\|^2_{\mathcal{HS}^n(H)} < +\infty,$$

called the Bargmann-Segal-Dwyer space. Members of $\mathcal{F}^r(H)$ are called Bargmann-Segal analytic functionals. See also [24].

3.3. Test and generalized white noise functionals

To study nonlinear functionals of white noise, Hida originally established a test-generalized functions setting $(L^2)^+ \subset (L^2) \subset (L^2)^-$ (see [14, 16, 38]). After Hida, Kubo and Takenaka [17] reformulated Hida's theory by taking different setting $(\mathcal{S}) \subset (L^2) \subset (\mathcal{S})'$. The space $(\mathcal{S})$ is an infinite dimensional analogue of the Schwartz space $\mathcal{S}$ on $\mathbb{R}$. We briefly describe as follows.

For $p \in \mathbb{R}$ and $\varphi \in (L^2)$, define

$$\|\varphi\|^2_{2,p} = \sum_{n=0}^{\infty} \frac{1}{n!} \|D^n S\varphi(0)\|^2_{\mathcal{HS}^n(\mathcal{S}_{-p})}. \tag{3.2}$$

and let $(\mathcal{S}_p)$ be the completion of the collection $\{\varphi \in (L^2); \|\varphi\|_{2,p} < +\infty\}$ with respect to $\| \cdot \|_{2,p}$-norm. Then $(\mathcal{S}_p)$, $p \in \mathbb{R}$, is a Hilbert space with the inner product induced by $\| \cdot \|_{2,p}$-norm. For $p, q \in \mathbb{R}$ with $q \geq p$, $(\mathcal{S}_q) \subset (\mathcal{S}_p)$ and the embedding $(\mathcal{S}_q) \hookrightarrow (\mathcal{S}_p)$ is of Hilbert-Schmidt type, whenever $q - p > 1/2$. Set $(\mathcal{S}) = \bigcap_{p>0} (\mathcal{S}_p)$ endowed with the projective limit topology. Then $(\mathcal{S})$ is a nuclear space and will serve as the space of test white noise functionals. The dual $(\mathcal{S})'$ of $(\mathcal{S})$ is the space of generalized white noise functionals (or often called Hida distributions). By identifying the dual

$(\mathcal{S}_p)'$ of $(\mathcal{S}_p)$, $p > 0$, with $(\mathcal{S}_{-p})$, we have a Gel'fand triple $(\mathcal{S}) \subset (L^2) \subset (\mathcal{S})'$ and the continuous inclusions: for $p \geq q > 0$,

$$(\mathcal{S}) \subset (\mathcal{S}_p) \subset (\mathcal{S}_q) \subset (L^2) \subset (\mathcal{S}_{-q}) \subset (\mathcal{S}_{-p}) \subset (\mathcal{S})',$$

where $(\mathcal{S})'$ is the inductive limit of the $(\mathcal{S}_{-p})$, $p > 0$. Hereafter, the dual pairing of $(\mathcal{S})'$ and $(\mathcal{S})$ will be denoted by $\langle\!\langle \cdot, \cdot \rangle\!\rangle$. One notes that $(\mathcal{S}_p)$, $p \geq 0$, is the domain of the second quantization $\Gamma(\mathbf{A}^p)$ of $\mathbf{A}^p$ and $(\mathcal{S}_0) = (L^2)$.

For $\varphi \in (\mathcal{S}_p)$ with $p > 0$, it is natural to extend the domain of $D^n S\varphi(0)$ to $\mathscr{C}\mathcal{S}_{-p} \times \cdots \times \mathscr{C}\mathcal{S}_{-p}$ and then define

$$S\varphi(z) = \sum_{n=0}^{\infty} \frac{1}{n!} D^n S\varphi(0) z^n, \quad z \in \mathscr{C}\mathcal{S}_{-p}.$$

It is clear that

$$\left| S\varphi(z) \right| \leq \|\varphi\|_{2,p} \cdot e^{\frac{1}{2}|z|^2_{-p}}, \quad z \in \mathscr{C}\mathcal{S}_{-p}.$$

On the other hand, by directly computing (3.2), it is easy to see that $e^{(\cdot, \eta)} \in (\mathcal{S}_p)$ for any $\eta \in \mathscr{C}\mathcal{S}_p$, $p > 0$. We then extend the S-transform to a function $F \in (\mathcal{S}_{-p})$, $p > 0$, by setting

$$SF(\eta) = e^{-\frac{1}{2}\int_{-\infty}^{\infty} \eta(t)^2 \, dt} \langle\!\langle F, e^{(\cdot, \eta)} \rangle\!\rangle, \quad \eta \in \mathscr{C}\mathcal{S}_p,$$

where $\|e^{-\frac{1}{2}\int_{-\infty}^{\infty} \eta(t)^2 \, dt + (\cdot, \eta)}\|_{2,p} = e^{\frac{1}{2}|\eta|^2_p}$. In fact, the S-transform is a unitary operator from $(\mathcal{S}_p)$ onto $\mathcal{F}^1(\mathscr{C}\mathcal{S}_{-p})$ for any $p \in \mathbb{R}$ (see [24]). In other words, for any $F \in (\mathcal{S}_p)$ with $p \in \mathbb{R}$,

$$\|F\|_{2,p}^2 = \sum_{n=0}^{\infty} \frac{1}{n!} \|D^n SF(0)\|^2_{\mathcal{HS}^n(\mathcal{S}_{-p})}.$$

Remark 3.2: The image SF, $F \in (\mathcal{S})'$, is also called the U-functional associated with F. Hida then studied the white noise calculus of generalized white noise functionals through their U-functionals. See [14, 15, 16, 38].

— *Analytic version of* $(\mathcal{S})$

For $p \in \mathbb{R}$, denote by $\mathcal{A}_p$ the space of analytic functions f defined on $\mathscr{C}\mathcal{S}_{-p}$ satisfying the exponential growth condition:

$$\|f\|_{\mathcal{A}_p} \equiv \sup\{|f(z)|e^{-\frac{1}{2}|z|^2_{-p}}; z \in \mathscr{C}\mathcal{S}_{-p}\} < +\infty.$$

Then $(\mathcal{A}_p, \|\cdot\|_{\mathcal{A}_p})$ is a Banach space and, by restriction, $\mathcal{A}_p$ is continuously embedded in $\mathcal{A}_q$ for $p > q$. Set $\mathcal{A}_\infty = \bigcap_{p \in \mathbb{R}} \mathcal{A}_p$ endowed with the projective limit topology. Then $\mathcal{A}_\infty$ becomes a locally convex topological algebra.

For any $\varphi \in (\mathcal{S}_p)$ with $p > \frac{1}{2}$, let

$$\widetilde{\varphi}(z) = \sum_{n=0}^{\infty} \frac{1}{n!} \int_{\mathcal{S}_{-p}} D^n S\varphi(0)(z + \mathrm{i}\,y)^n \, \mu(dy), \quad z \in \mathscr{C}\mathcal{S}_{-p},$$

which is well-defined since $\mathrm{supp}\,(\mu) \subset \mathcal{S}_{-p}$ with $p > \frac{1}{2}$. One notes that the above sum converges absolutely and uniformly on each bounded set in $\mathscr{C}\mathcal{S}_{-p}$. In addition, $\varphi = \widetilde{\varphi}$ almost all in $\mathcal{S}'$ with respect to μ.

Theorem 3.3: [24, 25, 26]

(i) Let $p > 1$, $r > 1$ and $s > \frac{1}{2}$. Then, for any $\varphi \in (\mathcal{S}_p)$, $\widetilde{\varphi}$ is analytic on $\mathscr{C}\mathcal{S}_{-p}$, and there are two constants $\alpha_p > 0$ and $\beta_{p,r} > 0$ such that

$$\alpha_p \cdot \|\widetilde{\varphi}\|_{\mathcal{A}_{p-s}} \leq \|\varphi\|_{2,p} \leq \beta_{p,r} \cdot \|\widetilde{\varphi}\|_{\mathcal{A}_{p+r}}.$$

(ii) Let $p > 0$ and $r > \frac{1}{2}$. Then, for any $\varphi \in (\mathcal{S}_{p+r})$,

$$C_r^{-1}\|\varphi\|_{2,p} \leq \|S\varphi\|_{\mathcal{A}_{p+r}} \leq \|\varphi\|_{2,p+r},$$

where $C_r^{3/2} = \int_{\mathcal{S}'} e^{|x|^2_{-r}} \, \mu(dx)$.

(iii) Let $p \in \mathbb{R}$. There exists $\alpha_p > 0$ such that

$$\alpha_p \cdot \|f\|_{\mathcal{F}^1(\mathscr{C}\mathcal{S}_{-p+2})} \leq \|f\|_{\mathcal{A}_p} \leq \|f\|_{\mathcal{F}^1(\mathscr{C}\mathcal{S}_{-p})}$$

for any analytic function f on $\mathscr{C}\mathcal{S}_{-p}$.

Corollary 3.4: [24]

(i) For $\varphi, \psi \in (\mathcal{S}_p)$ with $p > 1$, $\varphi(x) = \psi(x)$ μ-a.e. x in $\mathcal{S}'$ if and only if $\widetilde{\varphi}(z) = \widetilde{\psi}(z)$ for all $z \in \mathscr{C}\mathcal{S}_{-p}$.

(ii) $\mathcal{A}_\infty|_{\mathcal{S}'} \equiv \{\varphi|_{\mathcal{S}'}; \varphi \in \mathcal{A}_\infty\} \subset (\mathcal{S})$ and $\mathcal{A}_\infty = \widetilde{(\mathcal{S})} \equiv \{\widetilde{\varphi}; \varphi \in (\mathcal{S})\}$.

(iii) The families of norms $\{\|\cdot\|_{\mathcal{A}_p}; p > 0\}$ and $\{\|\cdot\|_{2,p}; p > 0\}$ are equivalent in $\mathcal{A}_\infty$.

Recall that any function $f \in (L^2)$ is identified with the equivalent class of functions in which any function is equal to f almost all with respect to μ. In this sense, we identify $(\mathcal{S})$ with $\mathcal{A}_\infty$, and call $\mathcal{A}_\infty$ the analytic version of $(\mathcal{S})$. It is noted that, for any $F \in (\mathcal{S})'$ and $\varphi \in (\mathcal{S})$, $\langle\!\langle F, \varphi \rangle\!\rangle = \langle\!\langle F, \widetilde{\varphi} \rangle\!\rangle$.

4. Hida Derivatives

For $F \in (\mathcal{S})'$ and $\eta \in \mathcal{S}$, define $\partial_\eta F \in (\mathcal{S})'$ by

$$\langle\!\langle \partial_\eta F, \varphi \rangle\!\rangle = \langle\!\langle F, \widetilde{\eta}\varphi \rangle\!\rangle - \langle\!\langle F, D_\eta \varphi \rangle\!\rangle, \quad \varphi \in (\mathcal{S}), \tag{4.1}$$

where $\widetilde{\eta}(x) = (x, \eta)$, $x \in \mathcal{S}'$, and $D_\eta \varphi$ is the Gâteaux derivative of φ in the direction of η.

Remark 4.1: By using the chain rule and applying the integration by parts formula given in Theorem 5.1,

$$\langle\!\langle D_\eta \varphi, \psi \rangle\!\rangle = \int_{\mathcal{S}'} D_\eta(\varphi\psi)(x)\,\mu(dx) - \int_{\mathcal{S}'} \varphi(x)\, D_\eta \psi(x)\,\mu(dx)$$

$$= \int_{\mathcal{S}'} (x, \eta)\, \varphi(x)\, \psi(x)\,\mu(dx) - \int_{\mathcal{S}'} \varphi(x)\, D_\eta \psi(x)\,\mu(dx)$$

for any φ, $\psi \in (\mathcal{S})$ and $\eta \in \mathcal{S}$. This implies that $\partial_\eta \varphi = D_\eta \varphi$.

Let $F \in (\mathcal{S}_p)$, $p \in \mathbb{R}$, and $\eta \in \mathcal{S}$. Putting $\varphi = e^{(\cdot, h) - \frac{1}{2}\int_{-\infty}^{\infty} h(t)^2\,dt}$, $h \in \mathscr{C}\mathcal{S}$, in (4.1) and applying the Cauchy integral formula,

$$S\partial_\eta F(h) = \frac{d}{dw}\bigg|_{w=0} SF(h + w\,\eta)$$

$$= \frac{1}{2\pi i} \int_{|w| = \frac{1}{2}|\eta|_{-p}^{-1}} \frac{SF(h + w\,\eta)}{w^2}\,dw, \qquad (4.2)$$

from which it follows that

$$|S\partial_\eta F(h)| \leq \text{Const.}\,\|F\|_{2,p} \cdot |\eta|_{-p} \cdot e^{\frac{1}{2}|h|^2_{-p+1}}. \qquad (4.3)$$

By virtue of (4.2), (4.3) and Theorem 3.3, we can conclude the following facts: For $F \in (\mathcal{S}_p)$ with $p \in \mathbb{R}$, $\partial_\eta F$ in (4.2) can be extended to $\eta \in \mathscr{C}\mathcal{S}_{-p}$ by defining

$$\partial_\eta F = S^{-1}\left(\frac{d}{dw}\bigg|_{w=0} SF(\cdot + w\eta)\right).$$

Moreover, by applying the characterization theorem in [27] (see also [33]),

$$\|\partial_\eta F\|_{2,p-3} \leq \text{Const.}\,\|F\|_{2,p} \cdot |\eta|_{-p}, \qquad (4.4)$$

where such a constant is independent of the choice of p, F, η. It is noted that $\partial_\eta \varphi = \partial_\eta \widetilde{\varphi}$ for $\varphi \in (\mathcal{S}_p)$ with $p > \frac{1}{2}$.

By (4.4), the mapping $(\eta, \varphi) \to \langle\!\langle \partial_\eta F, \varphi \rangle\!\rangle$, $F \in (\mathcal{S})'$, is bilinear and continuous from $(\mathcal{S}) \times \mathscr{C}\mathcal{S}$ into $\mathbb{C}$. By applying the kernel theorem (see [4]), there exists a unique element $K_F \in \mathscr{C}\mathcal{S}' \otimes (\mathcal{S})'$ such that

$$\langle\!\langle \partial_\eta F, \varphi \rangle\!\rangle = (\!(K_F, \eta \otimes \varphi)\!),$$

where $(\!(\cdot, \cdot)\!)$ is the $\mathscr{C}\mathcal{S} \otimes (\mathcal{S})'$–$\mathscr{C}\mathcal{S} \otimes (\mathcal{S})$ pairing. Symbolically, we express such an identity by the formal integral as follows:

$$\langle\!\langle \partial_\eta F, \varphi \rangle\!\rangle = \int_{-\infty}^{\infty} \int_{\mathcal{S}'} K_F(t; x)\, \eta(t)\, \varphi(x)\, \mu(dx) dt. \qquad (4.5)$$

$- \dot{B}(t)$-differentiation

Assume that $\frac{\ln 6}{2 \ln 2} < q < p$. Let $\varphi \in (\mathcal{S}_p)$ and $\eta \in \mathscr{C}\mathcal{S}_{-p}$. Observe that

$$\frac{d}{dw}\bigg|_{w=0} S\varphi(z + w\eta)$$

$$= \frac{d}{dw}\bigg|_{w=0} \sum_{n=0}^{\infty} \frac{1}{n!} D^n S\varphi(0)(z + w\eta)^n$$

$$= \sum_{n=1}^{\infty} \frac{1}{(n-1)!} D^n S\varphi(0) z^{n-1}\eta$$

$$= \sum_{n=1}^{\infty} \frac{1}{(n-1)!} \int_{\mathcal{S}_{-q}} \int_{\mathcal{S}_{-q}} D^n S\varphi(0)(x + z + iy)^{n-1}\eta \, \mu(dy)\mu(dx), \quad (4.6)$$

and

$$\sum_{n=1}^{\infty} \frac{1}{(n-1)!} \left| \int_{\mathcal{S}_{-q}} D^n S\varphi(0)(x + z + iy)^{n-1}\eta \, \mu(dy) \right|$$

$$\leq \left\{ \sum_{n=1}^{\infty} \frac{1}{(n-1)!} \|D^n S\varphi(0)\|_{\mathcal{HS}^n(\mathscr{C}\mathcal{S}_{-q})}^2 \right\}^{\frac{1}{2}} \cdot \int_{\mathcal{S}'} e^{3|y|^2_{-q}} \mu(dy)$$

$$\times |\eta|_{-q} \, e^{\frac{3}{2}|z|^2_{-q}} e^{\frac{3}{2}|x|^2_{-q}}$$

$$\leq \omega_{p-q} \cdot |\eta|_{-q} \, e^{\frac{3}{2}|z|^2_{-q}} e^{\frac{3}{2}|x|^2_{-q}} \|\varphi\|_{2,p} \int_{\mathcal{S}'} e^{3|y|^2_{-q}} \mu(dy) \ \in L^1(\mathcal{S}', \mu) \quad (4.7)$$

with respect to x for any $z \in \mathscr{C}\mathcal{S}_{-q}$. Here, we remark that the assumption "$q > \frac{\ln 6}{2 \ln 2}$" guarantees the integral $\int_{\mathcal{S}'} e^{3|y|^2_{-q}} \mu(dy)$ is finite. Therefore,

$$(3.6) = \int_{\mathcal{S}_{-q}} \left\{ \sum_{n=1}^{\infty} \frac{1}{(n-1)!} \int_{\mathcal{S}_{-q}} D^n S\varphi(0)(x + z + iy)^{n-1}\eta \, \mu(dy) \right\} \mu(dx)$$

$$= S \left(\sum_{n-1}^{\infty} \frac{1}{(n-1)!} : D^n S\varphi(0)x^{n-1}\eta : \right)(z), \quad z \in \mathscr{C}\mathcal{S}_{-q},$$

where

$$\left\| \sum_{n=1}^{\infty} \frac{1}{(n-1)!} : D^n S\varphi(0)x^{n-1}\eta : \right\|_{2,q}^2$$

$$= \sum_{n=1}^{\infty} \frac{1}{(n-1)!} \|D^n S\varphi(0)(\cdot, \ldots, \cdot, \eta)\|_{\mathcal{HS}(\mathscr{C}\mathcal{S}_{-q})}^2$$

$$\leq \omega_{p-q}^2 \cdot 2^{2(p-q)} \cdot \|\varphi\|_{2,p}^2 \cdot |\eta|_{-p}^2. \quad (4.8)$$

One notes that the inequality (4.8) is still valid for any $p \in \mathbb{R}$, $q < p$ and $\eta \in \mathscr{C}\mathscr{S}_{-p}$. Since $(\mathscr{S})$ and $\mathscr{C}\mathscr{S}$ are dense respectively in $(\mathscr{S}_p)$ and $\mathscr{C}\mathscr{S}_{-p}$ for any $p \in \mathbb{R}$, we can combine (4.8) with (4.4) and to extend the above argument to $\varphi \in (\mathscr{S})'$. In fact, we obtain the following

Proposition 4.2: *Let $\varphi \in (\mathscr{S}_p)$ with $p \in \mathbb{R}$ and $\eta \in \mathscr{C}\mathscr{S}_{-p}$.*

(i) For any $q < p$, $\partial_\eta \varphi \in (\mathscr{S}_q)$. In fact,

$$\partial_\eta \varphi = \sum_{n=1}^{\infty} \frac{1}{(n-1)!} : D^n S\varphi(0) x^{n-1} \eta :,$$

where

$$\|\partial_\eta \varphi\|_{2,q} \leq \omega_{p-q} \cdot 2^{p-q} \cdot \|\varphi\|_{2,p} \cdot |\eta|_{-p}.$$

(ii) For $\frac{\ln 6}{2\ln 2} < q < p$, the sum in (i) converges absolutely and uniformly on each bounded set in $\mathscr{C}\mathscr{S}_{-q}$, and $\partial_\eta \varphi = D_\eta \tilde{\varphi}$.

We commonly denote ∂_{δ_t} by ∂_t for $t \in \mathbb{R}$, where δ_t is the Dirac measure concentrated on t. If $\partial_t \varphi$ exists, that is, there is a $F \in (\mathscr{S})'$ such that its U-functional satisfies $SF = \frac{d}{dw}\big|_{w=0} S\varphi(\cdot + w\,\delta_t)$, φ is said to be $\dot{B}(t)$-differentiable and F is denoted by $\partial_t \varphi$. The operator ∂_t, which is sometimes written as $\dfrac{\partial}{\partial \dot{B}(t)}$, is often called the Hida derivative.

By a similar argument to (4.7), we get the following estimation: Assume that $\frac{\ln 6}{2\ln 2} < q < p$. Let $\varphi \in (\mathscr{S}_p)$ and $\eta \in \mathscr{C}\mathscr{S}_0$. For $z \in \mathscr{C}\mathscr{S}_{-q}$,

$$\sum_{n=1}^{\infty} \frac{1}{(n-1)!} \int_{-\infty}^{\infty} \int_{\mathscr{S}_{-p}} |\eta(t)||(\delta_t, h_n)||D^n S\varphi(0)(z+\mathrm{i}y)^{n-1} h_n|\, \mu(dy)\, dt$$

$$\leq \omega_{p-q} \cdot \|\varphi\|_{2,p} \cdot |\eta|_0 \cdot e^{|z|^2_{-q}} \left\{ \sum_{n=0}^{\infty} (2n+2)^{-2q} \right\}^{\frac{1}{2}} \int_{\mathscr{S}'} e^{|y|^2_{-q}}\, \mu(dy).$$

Then, for such φ, η and z,

$$\partial_\eta \varphi(z)$$

$$= \sum_{n=1}^\infty \frac{1}{(n-1)!} \int_{\mathcal{S}_{-q}} D^n S\varphi(0)(z+\mathrm{i}\,y)^{n-1}\eta \; \mu(dy)$$

$$= \sum_{k=0}^\infty (\eta, h_k) \sum_{n=1}^\infty \frac{1}{(n-1)!} \int_{\mathcal{S}_{-q}} D^n S\varphi(0)(z+\mathrm{i}\,y)^{n-1}h_k \; \mu(dy)$$

$$= \int_{-\infty}^\infty \eta(t) \sum_{n=1}^\infty \frac{1}{(n-1)!} \int_{\mathcal{S}_{-q}} \sum_{k=0}^\infty (\delta_t, h_k) \, D^n S\varphi(0)(z+\mathrm{i}\,y)^{n-1}h_k \; \mu(dy) \, dt$$

$$= \int_{-\infty}^\infty \eta(t) \sum_{n=1}^\infty \frac{1}{(n-1)!} \int_{\mathcal{S}_{-q}} D^n S\varphi(0)(z+\mathrm{i}\,y)^{n-1}\delta_t \; \mu(dy) \, dt$$

$$= \int_{-\infty}^\infty \eta(t) \, \partial_t\varphi(z) \, dt,$$

where we have by Proposition 4.2 that

$$\int_{-\infty}^\infty |\eta(t)| \, \|\partial_t\varphi\|_{2,q} \, dt \leq \omega_{p-q} \cdot 2^{p-q} \cdot \|\varphi\|_{2,p} \cdot |\eta|_0 \cdot \left\{ \sum_{n=0}^\infty (2n+2)^{-2p} \right\}^{\frac{1}{2}},$$

$$(4.9)$$

which is finite provided that $p > \frac{1}{2}$. Thus, by (4.9) and Proposition 4.2, we can also extend the above result to $\varphi \in (\mathcal{S}_p)$ with $p > \frac{1}{2}$ as follows.

Proposition 4.3:

(i) *For $\varphi \in (\mathcal{S}_p)$ with $p > \frac{1}{2}$ and $\eta \in \mathscr{C}\mathcal{S}_0$,*

$$\partial_\eta \, \varphi = \int_{-\infty}^\infty \eta(t) \, \partial_t\varphi \, dt \quad in \; (\mathcal{S}_q), \qquad (4.10)$$

for any $q < p$, where the right-hand integral exists in the sense of Bochner as an $(\mathcal{S}_q)$-valued integral satisfying the inequality (4.9).

(ii) *For $\frac{\ln 6}{2\ln 2} < q < p$, the formula (4.10) is valid pointwise in $\mathscr{C}\mathcal{S}_{-q}$:*

$$\partial_\eta \, \varphi(z) = D_\eta \, \varphi(z) = \int_{-\infty}^\infty \eta(t) \, D_{\delta_t}\varphi(z) \, dt \quad \forall \; z \in \mathscr{C}\mathcal{S}_{-q}.$$

Comparing (4.5) with Proposition 4.3 yields that $K_\varphi(t) = \partial_t \, \varphi$ for $\varphi \in (\mathcal{S}_p)$ with $p > \frac{1}{2}$, where K_φ is the kernel function given in (4.5). A question naturally arises:

"If $\varphi \in (L^2)$ and $\eta \in \mathscr{C}\mathcal{S}_0$, how about (4.10)?"

To see it, for any $n \in \mathbb{N}$, let $\phi_n \in \mathscr{C}L^2(\mathbb{R}^n, dt^{\otimes n})$ be a symmetric function such that

$$\int \cdots \int_{\mathbb{R}^n} \phi_n(t_1, \ldots, t_n)\, \eta_1 \widehat{\otimes} \cdots \widehat{\otimes} \eta_n(t_1, \ldots, t_n)\, dt_1 \cdots dt_n$$

$$= \frac{1}{n!} D^n S\varphi(0)(\eta_1, \ldots, \eta_n), \quad \eta_1, \ldots, \eta_n \in \mathscr{C}\mathcal{S}_0, \qquad (4.11)$$

where $\widehat{\otimes}$ means the symmetric tensor product. Then, for any $n \in \mathbb{N}$,

$$I_n(\phi_n) = \frac{1}{n!} : D^n S\varphi(0)x^n :,$$

where $I_n(\phi_n)$ is the multiple Wiener integral of order n with the kernel function ϕ_n. By the Fubini theorem,

$$\sum_{n=2}^{\infty} n! \int \cdots \int_{\mathbb{R}^{n-1}} |\phi_n(t, t_2, \ldots, t_n)|^2\, dt_2 \cdots dt_n < +\infty$$

for $[dt]$-almost all $t \in \mathbb{R}$. Observe that for $q < 0$,

$$\sum_{n=1}^{\infty} n^2(n-1)! \int \cdots \int_{\mathbb{R}^{n-1}} |(\mathbf{A}^q)^{\otimes(n-1)}\phi_n(t, t_2, \ldots, t_n)|^2\, dt_2 \cdots dt_n$$

$$\leq (1 + \omega_{-q}^2) \sum_{n=1}^{\infty} n! \int \cdots \int_{\mathbb{R}^{n-1}} |\phi_n(t, t_2, \ldots, t_n)|^2\, dt_2 \cdots dt_n. \qquad (4.12)$$

If $\varphi \in (\mathcal{S})$, it follows from (4.11) that $\phi_n \in \mathscr{C}\mathcal{S}^{\widehat{\otimes} n}$ and

$$\phi_n(t_1, \ldots, t_n) = \frac{1}{n!} D^n S\varphi(0)(\delta_{t_1}, \ldots, \delta_{t_n})$$

for any $t_1, \ldots, t_n \in \mathbb{R}$; moreover, for any $h \in \mathscr{C}\mathcal{S}$,

$$S(\partial_t\varphi)(h) = \sum_{n=1}^{\infty} \frac{1}{(n-1)!} D^n S\varphi(0)h^{n-1}\delta_t$$

$$= \sum_{n=1}^{\infty} n \int \cdots \int_{\mathbb{R}^{n-1}} \phi_n(t, t_2, \ldots, t_n)\, h(t_2) \cdots h(t_n)\, dt_2 \cdots dt_n$$

$$= S\left(\sum_{n=1}^{\infty} n\, I_{n-1}(\phi_n(t, \ldots))\right)(h), \qquad (4.13)$$

which implies that $\partial_t\varphi = \sum_{n=1}^{\infty} n\, I_{n-1}(\phi_n(t, \ldots))$, where it follows from (4.12) that for any $q < 0$,

$$\left\| \sum_{n=1}^{\infty} n\, I_{n-1}(\phi_n(t, \ldots)) \right\|_{2,q}^2 \leq (1 + \omega_{-q}^2) \sum_{n=1}^{\infty} n\, \|I_{n-1}(\phi_n(t, \ldots))\|_{2,0}^2. \quad (4.14)$$

Combine (4.13) with Proposition 4.3 and then extend φ to (L^2) by using (4.4) and (4.14). Then we have

Proposition 4.4: *For $\varphi \in (L^2)$ and $\eta \in \mathscr{C}S_0$,*

$$\partial_\eta \varphi = \int_{-\infty}^{\infty} \eta(t) \left\{ \sum_{n=1}^{\infty} n\, I_{n-1}(\phi_n(t, \ldots)) \right\} dt \quad in \ (S_q)$$

for any $q < 0$, where the right-hand integral exists in the sense of Bochner as an (S_q)-valued integral satisfying the inequality

$$\int_{-\infty}^{\infty} |\eta(t)| \left\| \sum_{n=1}^{\infty} n\, I_{n-1}(\phi_n(t, \ldots)) \right\|_{2,q} dt \leq \sqrt{1 + \omega_{-q}^2} \cdot |\eta|_0 \cdot \|\varphi\|_{2,0}.$$

Corollary 4.5: *For $\varphi \in (L^2)$, the kernel function $K_\varphi(t; \cdot)$ in (4.5) is exactly $\sum_{n=1}^{\infty} n\, I_{n-1}(\phi_n(t, \ldots))$, where, for any $q < 0$,*

$$\int_{-\infty}^{\infty} \|K_\varphi(t; \cdot)\|_{2,q}^2 \, dt \leq (1 + \omega_{-q}^2) \cdot \|\varphi\|_{2,0}^2.$$

5. Integration by Parts Formula

For any probability measure $\mathcal{P}$ on $(\mathcal{S}', \mathscr{B}(\mathcal{S}'))$, let $\mathcal{G}_\mathcal{P}$ be the class consisting of all complex-valued functions φ on $\mathcal{S}'$ satisfying the conditions: For any $h \in S_0$, (a) φ is Gâteaux differentiable at x in the direction of h for any $x \in \mathcal{S}'$; (b) both φ and $\delta\varphi(\cdot; h)$ belong to $\mathscr{C}L^\alpha(\mathcal{S}', \mathcal{P})$ for some $\alpha > 1$, where $\delta\varphi(x; h)$ the Gâteaux derivative of φ at $x \in \mathcal{S}'$ in the direction of h.

For any $h \in S_0$, let $e(h) = e^{\langle \cdot, h \rangle - \frac{1}{2} \int_{-\infty}^{\infty} |h(t)|^2 \, dt}$. Then, for $\varphi \in \mathcal{G}_\mu$,

$$\int_{\mathcal{S}'} \frac{\varphi(x + rh) - \varphi(x)}{r} \mu(dx) = \int_{\mathcal{S}'} \varphi(x) \cdot \left\{ \frac{e(rh)(x) - 1}{r} \right\} \mu(dx)$$

$$\to \int_{\mathcal{S}'} \langle x, h \rangle\, \varphi(x)\, \mu(dx), \quad as \ r \to 0^+,$$

where the last term is obtained by the inequality that

$$\left| \frac{e(rh)(x) - 1}{r} \right| \leq \left\{ |\langle x, h \rangle| + \int_{-\infty}^{\infty} |h(t)|^2 \, dt \right\} \cdot e^{|\langle x, h \rangle|} \in L^{\alpha'}(\mathcal{S}', \mu)$$

for any $0 < z < 1$, $\frac{1}{\alpha} + \frac{1}{\alpha'} = 1$, and then applying the dominated convergence argument. On the other hand,

$$\int_{S'} \frac{\varphi(x+rh) - \varphi(x)}{r} \mu(dx) = \int_{S'} \frac{1}{r} \int_0^r \delta\varphi(x + \vartheta\, h; \eta)\, d\vartheta\, \mu(dx)$$

$$= \int_{S'} \delta\varphi(x; h) \cdot \left\{ \frac{1}{r} \int_0^r e(\vartheta\, h)\, d\vartheta \right\} \mu(dx)$$

$$\to \int_{S'} \delta\varphi(x; h)\, \mu(dx), \quad \text{as } r \to 0^+,$$

where the last term is obtained by applying the dominated convergence argument. Putting together the above formulas shows the following integration by parts formula for the white noise measure μ.

Theorem 5.1: *Let $\varphi : S' \to \mathbb{C}$ be in the class $\mathcal{G}_\mu$. Then, for any $h \in S_0$,*

$$\int_{S'} \langle x, h \rangle\, \varphi(x)\, \mu(dx) = \int_{S'} \delta\varphi(x; h)\, \mu(dx). \qquad (5.1)$$

Remark 5.2: The integration by parts formula for abstract Wiener measures was obtained by Kuo [18] in 1974. Recently, under much weak conditions, Kuo and Lee [20] reformulated this formula and simplified the proof by applying the technique Stein used in proving his famous Stein's lemma (Proposition 2.1) for normal distribution (see [39, 40]).

The formula (5.1) is an infinite dimensional analogue of the Stein identity for normal distribution, which also characterizes the white noise measure as follows.

Theorem 5.3: *A probability measure $\mathcal{P}$ on $(S', \mathcal{B}(S'))$ is equal to the white noise measure μ if and only if for any $\varphi \in \mathcal{G}_\mathcal{P}$, the following equality holds:*

$$\int_{S'} \left\{ \langle x, \eta \rangle\, \varphi(x) - \delta\varphi(x; \eta) \right\} \mathcal{P}(dx) = 0, \quad \forall\, \eta \in S. \qquad (5.2)$$

Proof: If $\mathcal{P} = \mu$, then relation (5.2) is satisfied by virtue of Theorem 5.1. Now, suppose $\mathcal{P}$ satisfies (5.2). Let

$$\phi_\eta(r) = \int_{S'} e^{ir\langle x, \eta \rangle}\, \mathcal{P}(dx), \quad r \in \mathbb{R},\ \eta \in S.$$

By the mean value theorem for differentiation, there are two real numbers p_{sr}, q_{sr} between s and r such that

$$e^{i\, s\langle x, \eta \rangle} - e^{i\, r\langle x, \eta \rangle} = \frac{i\,(s-r)}{2} \cdot \langle x, \eta \rangle \cdot \Psi_{s,r;\eta}(x), \quad x \in S',$$

where $\Psi_{s,r;\eta}(x) = e^{i\,p_{sr}(x,\,\eta)} - e^{-i\,p_{sr}(x,\,\eta)} + e^{i\,q_{sr}(x,\,\eta)} + e^{-i\,q_{sr}(x,\,\eta)}$. Then

$$\frac{\phi_\eta(s) - \phi_\eta(r)}{s - r} = (i/2) \int_{S'} (x, \eta) \cdot \Psi_{s,r;\eta}(x) \, \mathcal{P}(dx).$$

It is clear that $\Psi_{s,r;\eta} \in \mathcal{G}_\mathcal{P}$ Then, by (5.2) we have

$$\int_{S'} (x, \eta) \cdot \Psi_{s,r;\eta}(x) \, \mathcal{P}(dx) = i \sum_{j=1}^{4} a_j v_j \, \phi_\eta(v_j) \, |\eta|_0^2,$$

where $v_1 = p_{sr} = -v_2$, $v_3 = q_{sr} = -v_4$, $a_1 = a_3 = a_4 = 1, a_2 = -1$. Letting s tend to r, we see that $\frac{d\phi_\eta(r)}{dr}$ exists, and $\frac{d\phi_\eta(r)}{dr} = -r\phi_\eta(r) \, |\eta|_0^2$ with $\phi_\eta(0) = 1$. Thus, $\phi_\eta(r) = e^{-\frac{1}{2}r^2 |\eta|_0^2}$ and $\mathcal{P} = \mu$. $\qquad\square$

By Theorem 5.3 and an observation of its proof, we can recover the following version of Stein's characterization of the normal distribution.

Corollary 5.4: (*Stein's lemma*) *A real-valued random variable Y has the standard normal distribution if and only if*

$$\mathbb{E}[Y f(Y)] = \mathbb{E}[f'(Y)] \qquad (5.3)$$

for any bounded complex-valued function f with bounded derivative f'.

Let f be a function defined on S' with values in a complex Banach space W. Then f is said to be S_0-differentiable if the mapping $\phi(h) \equiv f(x + h)$, $h \in S_0$, is Fréchet differentiable at 0 for any $x \in S'$. The Fréchet derivative $\phi'(0)$ at $0 \in S_0$ is called the S_0-derivative of f at $x \in S'$, denoted by $\langle Df(x), h \rangle$. The k-th order S_0-derivatives of f at x are defined inductively and denoted by $D^k f(x)$ for $k \geq 2$ if they exist. One notes that $D^k f(x)$ is a bounded k-linear mapping from the Cartesian product $S_0 \times \cdots \times S_0$ of k copies of S_0 into W for any $k \in \mathbb{N}$. In particular, when $W = \mathbb{R}$, $Df(x) \in S_0$ and $D^2 f(x)$ is regarded as a bounded linear operator from S_0 into S_0 for any $x \in S'$ (see [19]).

We can use Theorem 5.1 to obtain another integration by parts formula which also characterizes the white noise measures.

Theorem 5.5: (*cf.* [20]) *A probability measure $\mathcal{P}$ on $(S', \mathscr{B}(S'))$ is equal to the white noise measure μ if and only if, for any S_0-differentiable function f on S' such that $f(x) \in \mathscr{C}S_p$, $p > \frac{1}{2}$, for any $x \in S'$, $\|Df(\cdot)\|_{\mathrm{tr}} \in L^1(S', \mathcal{P})$ and $\int_{S'} |f(x)|_p^\alpha \, \mathcal{P}(dx) < \infty$ for some $\alpha > 1$, the following equality holds:*

$$\int_{S_{-p}} (x, f(x))_p \, \mathcal{P}(dx) = \int_{S'} \mathrm{Tr}(Df(x)) \, \mathcal{P}(dx), \qquad (5.4)$$

where $(\cdot, \cdot)_p$ is the $\mathscr{C}S_{-p}$-$\mathscr{C}S_p$ pairing, $\mathrm{Tr}(\cdot)$ denotes the trace of a trace class operator on S_0 and $\|\cdot\|_{\mathrm{tr}}$ means the trace class norm.

Proof: *Necessity.* Let $\{h_n\}_{n=0}^{\infty}$ be the CONS for S_0 as mentioned in Subsection 3.1, and for any $x \in S'$ and $n \in \mathbb{N}_0$, let $P_n x = \sum_{j=0}^{n} (x, h_j) h_j$. Since f is S_0-differentiable, $r^{-1}(f(x+rh) - f(x))$ converges to $\langle Df(x), h \rangle$ uniformly with respect to h on each bounded set in S_0 as $r \to 0$ for any $x \in S'$. This implies that $\varphi(x) \equiv (f(x), h_j)$, $x \in S'$, is Gâteaux differentiable at x in the direction of $h \in S_0$, and $\delta\varphi(x; h) = \langle\langle Df(x), h\rangle, h_j\rangle_0$ for any $j \in \mathbb{N}_0$. Moreover, it follows from the conditions (a) and (b) and by applying the Fernique theorem (see [19]) that φ is in $\mathcal{G}_\mu$. Then, by applying Theorem 5.1, we see that

$$\int_{S_{-p}} (P_n x, f(x))_p \, \mu(dx) = \sum_{j=0}^{n} \int_{S_p} (f(x), h_j)(x, h_j) \, \mu(dx)$$

$$= \sum_{j=0}^{n} \int_{S_{-p}} \langle\langle Df(x), h_j\rangle, h_j\rangle_0 \, \mu(dx)$$

$$= \int_{S_{-p}} \mathrm{Tr}(\langle Df(x), P_n(\cdot)\rangle) \, \mu(dx). \qquad (5.5)$$

Note that for all $x \in S_{-p}$ and $n \in \mathbb{N}_0$,

$$\left| \mathrm{Tr}(\langle Df(x), P_n(\cdot)\rangle) \right| \leq \|Df(x)\|_{\mathrm{tr}},$$

$$\left| (P_n x, f(x)) \right| \leq |f(x)|_p \, |P_n x|_{-p} \leq |f(x)|_p \, |x|_{-p}.$$

Let n tend to infinity, and then obtain Equation (5.4) by applying the Lebesgue dominated convergence theorem to (5.5).

Sufficiency. Fix $\eta \in S$. Let $f(x) = \mathrm{i}\, e^{\mathrm{i}(x, \eta)} \eta$ for any $x \in S'$. Then f is Fréchet differentiable on S', and $\langle Df(x), y\rangle = -(y, \eta)\, e^{\mathrm{i}(x, \eta)}\eta$ for any $y \in S'$. Thus $Df(x)$ is a bounded linear operator from S_{-p} into $\mathscr{C}S_p$ for $p > \frac{1}{2}$, the operator norm of which is less than or equal to $|\eta|_p^2$. By applying the Goodman theorem and Fernique theorem (see [19]),

$$\int_{S'} \|Df(x)\|_{\mathrm{tr}} \, \mathcal{P}(dx) \leq |\eta|_p^2 \int_{S_{-p}} |x|_{-p}^2 \, \mu(dx) < +\infty.$$

In addition, for any $1 \leq \alpha < +\infty$, $\int_{S'} |f(x)|_p^\alpha \, \mathcal{P}(dx) = |\eta|_p^\alpha < +\infty$. By the assumption, f satisfies the identity (5.4) and we have

$$\int_{S'} (x, \eta)\, e^{\mathrm{i}(x, \eta)} \, \mathcal{P}(dx) = \mathrm{i}\, |\eta|_0^2 \int_{S'} e^{\mathrm{i}(x, \eta)} \, \mathcal{P}(dx).$$

By the same argument as in the proof of Theorem 5.3, $\int_{\mathcal{S}'} e^{\mathrm{i}\,(x,\,\eta)}\,\mathcal{P}(dx) = e^{-\frac{1}{2}|\eta|_0^2}$. The proof is complete. $\qquad\qquad\qquad\qquad\qquad\qquad\qquad\qquad\square$

Remark 5.6: In Theorem 5.5, the assumption "$p > \frac{1}{2}$" is necessary based on the fact that $(\mathcal{S}_0, \mathcal{S}_{-p})$, $p > \frac{1}{2}$, is an abstract Wiener space.

– *Application to Number operators*

Let f be a complex-valued function on $\mathcal{S}'$. If f is twice $\mathcal{S}_0$-differentiable at $x \in \mathcal{S}'$ and $D^2 f(x)$ is a trace-class operator on $\mathcal{S}_0$, its trace is known as the Gross Laplacian $\Delta_{_G} f(x)$ of f at x: $\Delta_{_G} f(x) = \mathrm{Tr}(D^2 f(x))$ (see [13]). In particular, if f is twice Fréchet differentiable in $\mathcal{S}_{-p}$ with $p > \frac{1}{2}$, then the restriction $D^2 f(x)|_{\mathcal{S}_0}$ of $D^2 f(x)$ to $\mathcal{S}_0$ is automatically of trace class on $\mathcal{S}_0$ by the Goodman theorem (see [19]).

Now, if f is twice $\mathcal{S}_0$-differentiable at $x \in \mathcal{S}_{-p}, p > \frac{1}{2}$, such that $Df(x) \in \mathcal{S}_p$ and $D^2 f(x)$ is a trace-class operator on $\mathcal{S}_0$, we define the Beltrami Laplacian

$$\Delta_{_B} f(x) = \Delta_{_G} f(x) - (x,\, Df(x))_p.$$

For $\varphi \in (\mathcal{S})$ and $x \in \mathcal{S}'$, $D\varphi(x)$ is a continuous linear functional on $\mathcal{S}'$. Since $\mathcal{S}$ is a nuclear space, $D\varphi(x) \in \mathcal{S}$. Similarly, $D^2\varphi(x)$ is a continuous linear operator from $\mathcal{S}'$ into $\mathcal{S}$. Then it follows from the Goodman theorem that $D^2\varphi(x)|_{\mathcal{S}_0}$ is a trace-class operator on $\mathcal{S}_0$, and thus $\Delta_{_B}\varphi(x)$ exists.

As a consequence of Theorem 5.5, we can see that for any $n \in \mathbb{N}_0$,

$$\Delta_{_B} \int_{\mathcal{S}'} \prod_{j=1}^{n}(x+\mathrm{i}\,y,\, h_{k_j})\,\mu(dy) = -n \int_{\mathcal{S}'} \prod_{j=1}^{n}(x+\mathrm{i}\,y,\, h_{k_j})\,\mu(dy),$$

where k_j's $\in \mathbb{N}_0$. See also [21]. In fact, $\Delta_{_B}$ is densely defined on (L^2). The closure of the operator $-\Delta_{_B}$, denoted by $\mathcal{N}$, is known as the number operator. Then the domain $\mathrm{Dom}(\mathcal{N})$ of $\mathcal{N}$ is

$$\mathrm{Dom}(\mathcal{N}) = \left\{ \varphi \in (L^2);\ \sum_{n=0}^{\infty} \frac{n^2}{n!} \|D^n S\varphi(0)\|_{\mathcal{H}S^n(\mathscr{C}\mathcal{S}_0)}^2 < +\infty \right\}.$$

It is obvious that $(\mathcal{S}_p) \subset \mathrm{Dom}(\mathcal{N})$ for $p \geq \frac{1}{2}$.

Now, let $\varphi, \psi \in (\mathcal{S})$. By Theorem 5.5,

$$\int_{\mathcal{S}'} \varphi(x) \, \mathcal{N}\psi(x) \, \mu(dx)$$

$$= -\int_{\mathcal{S}'} \varphi(x) \, \Delta_G \psi(x) \, \mu(dx) + \int_{\mathcal{S}'} \varphi(x) \, (x, \, D\psi(x)) \, \mu(dx)$$

$$= -\int_{\mathcal{S}'} \varphi(x) \, \Delta_G \psi(x) \, \mu(dx) + \int_{\mathcal{S}'} (x, \, \varphi(x) D\psi(x)) \, \mu(dx)$$

$$= -\int_{\mathcal{S}'} \varphi(x) \, \Delta_G \psi(x) \, \mu(dx) + \int_{\mathcal{S}'} \mathrm{Tr}(D(\varphi(\cdot)D\psi(\cdot))(x)) \, \mu(dx). \quad (5.6)$$

Observe that for any $y, z \in \mathcal{S}'$,

$$(D_y(\varphi(\cdot)D\psi(\cdot))(x), \, z) = D_y\varphi(x) \, D_z\psi(x) + \varphi(x) \, D^2\psi(x)(y, z)$$

$$= \partial_y\varphi(x) \, \partial_z\psi(x) + \varphi(x) \, D^2\psi(x)(y, z).$$

Then, for any $x \in \mathcal{S}'$, it follows from Proposition 4.3 that

$$\mathrm{Tr}(D(\varphi(\cdot)D\psi(\cdot))(x))$$

$$= \sum_{n=0}^{\infty} \partial_{h_n}\varphi(x) \, \partial_{h_n}\psi(x) + \varphi(x) \, \Delta_G\psi(x)$$

$$= \sum_{n=0}^{\infty} \int_{-\infty}^{\infty} h_n(t) \, \partial_t\varphi(x) \, dt \int_{-\infty}^{\infty} h_n(t) \, \partial_t\psi(x) \, dt + \varphi(x) \, \Delta_G\psi(x)$$

$$= \int_{-\infty}^{\infty} \partial_t\varphi(x) \, \partial_t\psi(x) \, dt + \varphi(x) \, \Delta_G\psi(x). \quad (5.7)$$

Combining (5.6) with (5.7), applying Proposition 4.2 and by extension, we have the following

Theorem 5.7: *For any $\varphi, \psi \in (\mathcal{S}_p)$ with $p > \frac{1}{2}$,*

$$\langle\!\langle \mathcal{N}\varphi, \, \psi \rangle\!\rangle_{2,0} = \int_{-\infty}^{\infty} \langle\!\langle \partial_t\varphi, \, \partial_t\psi \rangle\!\rangle_{2,0} \, dt,$$

where $\langle\!\langle \cdot, \cdot \rangle\!\rangle_{2,0}$ is the inner product induced by $\| \cdot \|_{2,0}$-norm. Moreover,

$$\int_{-\infty}^{\infty} \|\partial_t \, \varphi\|_{2,0} \|\partial_t \, \psi\|_{2,0} \, dt \leq \omega_p^2 \cdot 2^{2p} \cdot \|\varphi\|_{2,p} \cdot \|\psi\|_{2,p} \left\{ \sum_{n=0}^{\infty} (2n+2)^{-2p} \right\}.$$

Observe that for $\varphi \in (L^2)$, it is analogous to Corollary 3.5 that we have

$$\int_{-\infty}^{\infty} \|K_\varphi(t; \cdot)\|_{2,0}^2 \, dt = \sum_{n=1}^{\infty} n \, \|I_n(\phi_n)\|_{2,0}^2.$$

Then we have more general results than those in Proposition 4.4, Corollary 4.5 and Theorem 5.7 as follows.

Theorem 5.8:

(i) *Let* $\varphi \in \mathrm{Dom}(\mathcal{N}^{1/2})$. *Then* $K_\varphi(t) \in (L^2)$ *for* $[dt]$-*almost all* $t \in \mathbb{R}$ *and satisfies*

$$\int_{-\infty}^{\infty} \|K_\varphi(t; \cdot)\|_{2,0}^2 \, dt = \|\mathcal{N}^{1/2}\varphi\|_{2,0}^2.$$

Moreover, for any $\eta \in \mathscr{C}\mathcal{S}_0$,

$$\partial_\eta \varphi = \int_{-\infty}^{\infty} \eta(t) \, K_\varphi(t; \cdot) \, dt \quad in \ (L^2),$$

where

$$\|\partial_\eta \varphi\|_{2,0} \le |\eta|_0 \cdot \|\mathcal{N}^{1/2}\varphi\|_{2,0}.$$

(ii) *Let* $\varphi \in \mathrm{Dom}(\mathcal{N})$ *and* $\psi \in (L^2)$. *Then the sum*

$$\sum_{n=1}^{\infty} n^2 \langle\!\langle I_{n-1}(\phi_n(t, \dots)), \, I_{n-1}(\psi_n(t, \dots)) \rangle\!\rangle_{2,0},$$

denoted by $[K_\varphi(t; \cdot), K_\psi(t; \cdot)]_{2,0}$, *absolutely converges for* $[dt]$-*almost all* $t \in \mathbb{R}$, *where* ψ_n *is the kernel function defined analogously as* ϕ_n *in (4.11) by replacing* φ *by* ψ. *Moreover,*

$$\langle\!\langle \mathcal{N}\varphi, \psi \rangle\!\rangle_{2,0} = \int_{-\infty}^{\infty} [K_\varphi(t; \cdot), K_\psi(t; \cdot)]_{2,0} \, dt.$$

Note. In the sequel, we will identify $K_\varphi(t; \cdot)$ with $\partial_t \varphi$ for $\varphi \in (L^2)$.

Remark 5.9: If both $K_\varphi(t; \cdot)$ and $K_\psi(t; \cdot)$ are in (L^2), then

$$[K_\varphi(t; \cdot), K_\psi(t; \cdot)]_{2,0} = \langle\!\langle K_\varphi(t; \cdot), K_\psi(t; \cdot) \rangle\!\rangle_{2,0}.$$

In fact, $\mathrm{Dom}(\mathcal{N}^{1/2}) = \mathbb{D}^{1,2}$, and for $\varphi \in \mathrm{Dom}(\mathcal{N}^{1/2})$, $K_\varphi(\cdot, x)$ coincides with the Malliavin derivative $D\varphi(x)$ of φ, $x \in \mathcal{S}'$.

6. Connecting Stein's Method with Hida Calculus for White Noise Functionals

Let $\mathscr{H}$ be a separating class of Borel-measurable complex-valued test functions on $\mathbb{R}$, which means that any two real-valued random variables F, G satisfying $\mathbb{E}[h(F)] = \mathbb{E}[h(G)]$ for every $h \in \mathscr{H}$ have the same law. Of course F and G are assumed to be such that both $\mathbb{E}h(F)$ and $\mathbb{E}h(G)$ exists for all $h \in \mathscr{H}$. For any two such real-valued random variables F and G, the distance between the laws of F and G, induced by $\mathscr{H}$, is given by

$$d_{\mathscr{H}}(F, G) = \sup\{|\mathbb{E}[h(F)] - \mathbb{E}[h(G)]|;\ h \in \mathscr{H}\}.$$

Let $\mathscr{H}$ be such that $\mathbb{E}h(Z)$ exists for $h \in \mathscr{H}$, where Z has the standard normal distribution. Let f_h be the solution, given by (2.1), of the Stein equation

$$f'(w) - wf(w) = h(w) - \mathbb{E}h(Z). \tag{6.1}$$

Assume that $\mathscr{H}$ is such that f_h is bounded and absolutely continuous with bounded f'_h for $h \in \mathscr{H}$. Then for any random variable Y such that $\mathbb{E}h(Y)$ exists for $h \in \mathscr{H}$, we have

$$d_{\mathscr{H}}(Y, Z) = \sup\{|\mathbb{E}[f'_h(Y) - Yf_h(Y)]|;\ h \in \mathscr{H}\}.$$

Consider $(\mathcal{S}', \mathscr{B}(\mathcal{S}'), \mu)$ as the underlying probability space. Assume that $\varphi \in \mathrm{Dom}(\mathcal{N}^{1/2})$ with $\mathbb{E}[\varphi] = 0$. Define

$$\mathcal{N}^{-1}\varphi = \sum_{n=1}^{\infty} \frac{1}{nn!} : D^n S\varphi(0)x^n : .$$

Then $\varphi = \mathcal{N}\mathcal{N}^{-1}\varphi$ since $\mathbb{E}[\varphi] = 0$. It is noted that $\mathcal{N}^{-1}\varphi \in \mathrm{Dom}(\mathcal{N})$ and $f_h(\varphi) \in (L^2)$. By Theorem 5.8,

$$\int_{\mathcal{S}'} \varphi(x)\,f_h(\varphi(x))\,\mu(dx) = \langle\!\langle \mathcal{N}\mathcal{N}^{-1}\varphi,\ f_h(\varphi)\rangle\!\rangle_{2,0}$$

$$= \int_{-\infty}^{\infty} [\,\partial_t \mathcal{N}^{-1}\varphi,\ \partial_t f_h(\varphi)]_{2,0}\ dt. \tag{6.2}$$

Take a sequence $\{\varphi_k\} \subset (\mathcal{S})$ such that $\mathcal{N}^{1/2}\varphi_k \to \mathcal{N}^{1/2}\varphi$ in (L^2). Then

$$\|f_h(\varphi_k) - f_h(\varphi)\|_{2,0} \le |f'_h|_\infty \cdot \|\varphi_k - \varphi\|_{2,0} \to 0, \quad \text{as } k \to \infty,$$

and thus

$$\langle\!\langle \mathcal{N}\mathcal{N}^{-1}\varphi,\ f_h(\varphi_k)\rangle\!\rangle_{2,0} \to \langle\!\langle \mathcal{N}\mathcal{N}^{-1}\varphi,\ f_h(\varphi)\rangle\!\rangle_{2,0} \quad \text{as } k \to \infty. \tag{6.3}$$

On the other hand, for each $k \in \mathbb{N}$, it follows by the chain rule that

$$\frac{d}{dw}\Big|_{w=0} Sf_h(\varphi_k)(z + w\delta_t) = \frac{d}{dw}\Big|_{w=0} \int_{\mathcal{S}'} f_h(\varphi_k)(z + w\delta_t + x)\,\mu(dx)$$

$$= \int_{\mathcal{S}'} f'_h(\varphi_k(z + x))\,\partial_t\varphi_k(z + x)\,\mu(dx)$$

$$= S(f'_h(\varphi_k)\,\partial_t\varphi_k)(z), \quad z \in \mathscr{C}\mathcal{S}_0,$$

and then

$$\int_{-\infty}^{\infty} [\partial_t \mathcal{N}^{-1}\varphi, \partial_t f_h(\varphi_k)]_{2,0} \, dt = \int_{-\infty}^{\infty} \langle\!\langle \partial_t \mathcal{N}^{-1}\varphi, f_h'(\varphi_k) \partial_t \varphi_k \rangle\!\rangle_{2,0} \, dt.$$
(6.4)

By Theorem 5.8, $f_h'(\varphi) \partial_t \varphi \in (L^2)$ for almost all $t \in \mathbb{R}$. We would like to estimate

$$\left| \int_{-\infty}^{\infty} \langle\!\langle \partial_t \mathcal{N}^{-1}\varphi, f_h'(\varphi_k) \partial_t \varphi_k - f_h'(\varphi) \partial_t \varphi \rangle\!\rangle_{2,0} \, dt \right|$$

$$\leq \int_{-\infty}^{\infty} |\langle\!\langle \partial_t \mathcal{N}^{-1}\varphi, (f_h'(\varphi_k) - f_h'(\varphi)) \partial_t \varphi_k \rangle\!\rangle_{2,0}| \, dt$$

$$+ \int_{-\infty}^{\infty} |\langle\!\langle \partial_t \mathcal{N}^{-1}\varphi, f_h'(\varphi)(\partial_t \varphi_k - \partial_t \varphi) \rangle\!\rangle_{2,0}| \, dt$$

$$\leq 3|f_h'|_\infty \int_{-\infty}^{\infty} \langle\!\langle |\partial_t \mathcal{N}^{-1}\varphi|, |\partial_t \varphi_k - \partial_t \varphi| \rangle\!\rangle_{2,0} \, dt \tag{6.5}$$

$$+ \int_{-\infty}^{\infty} \langle\!\langle |\partial_t \mathcal{N}^{-1}\varphi|, |f_h'(\varphi_k) - f_h'(\varphi)| \, |\partial_t \varphi| \rangle\!\rangle_{2,0} \, dt. \tag{6.6}$$

By Theorem 5.8,

$$\int_{-\infty}^{\infty} \|\partial_t \varphi_k - \partial_t \varphi\|_{2,0}^2 \, dt = \|\mathcal{N}^{1/2}(\varphi_k - \varphi)\|_{2,0}^2 \to 0.$$

Then

$$(6.5) \leq 3|f_h'|_\infty \left\{ \int_{-\infty}^{\infty} \|\partial_t \mathcal{N}^{-1}\varphi\|_{2,0}^2 \, dt \right\}^{\frac{1}{2}} \left\{ \int_{-\infty}^{\infty} \|\partial_t \varphi_k - \partial_t \varphi\|_{2,0}^2 \, dt \right\}^{\frac{1}{2}} \to 0$$

as $k \to \infty$. In addition, since $\|\varphi_k \to \varphi\|_{2,0} \to 0$, there exists a subsequence, still written as $\{\varphi_k\}$, such that

$$\varphi_k(x) \to \varphi(x), \quad \text{for } [\mu]\text{-almost all } x \in \mathcal{S}'.$$

Hence

$$\Phi_k(t,x) := |\partial_t \mathcal{N}^{-1}\varphi(x)| \, |f_h'(\varphi_k(x)) - f_h'(\varphi(x))| \, |\partial_t \varphi(x)| \to 0$$

for $[dt \otimes \mu]$-almost all (t,x) in $\mathbb{R} \times \mathcal{S}'$, where

$$|\Phi_k(t,x)| \leq 2|f_h'|_\infty |\partial_t \mathcal{N}^{-1}\varphi(x)| \, |\partial_t \varphi(x)| \in L^1(\mathbb{R} \times \mathcal{S}', dt \otimes \mu).$$

Then it follows by the Lebesgue dominated convergence theorem that

(6.6)$\to$ 0 as $k \to \infty$. So we can conclude by (6.3)–(6.6) that

$$
\begin{aligned}
\langle\!\langle \mathcal{N}\mathcal{N}^{-1}\varphi, f_h(\varphi) \rangle\!\rangle_{2,0} &= \lim_{k\to\infty} \langle\!\langle \mathcal{N}\mathcal{N}^{-1}\varphi, f_h(\varphi_k) \rangle\!\rangle_{2,0} \\
&= \lim_{k\to\infty} \int_{-\infty}^{\infty} \langle\!\langle \partial_t\,\mathcal{N}^{-1}\varphi, \partial_t f_h(\varphi_k) \rangle\!\rangle_{2,0}\, dt \\
&= \lim_{k\to\infty} \int_{-\infty}^{\infty} \langle\!\langle \partial_t\,\mathcal{N}^{-1}\varphi, f_h'(\varphi_k)\,\partial_t\varphi_k \rangle\!\rangle_{2,0}\, dt \\
&= \int_{-\infty}^{\infty} \langle\!\langle \partial_t\,\mathcal{N}^{-1}\varphi, f_h'(\varphi)\,\partial_t\varphi \rangle\!\rangle_{2,0}\, dt \\
&= \int_{\mathcal{S}'} f_h'(\varphi(x)) \int_{-\infty}^{\infty} \partial_t\,\mathcal{N}^{-1}\varphi(x)\,\partial_t\varphi(x)\, dt\, \mu(dx).
\end{aligned}
$$

Together with (6.2) and $d_{\mathscr{H}}(\varphi, Z)$, we obtain

Theorem 6.1: *Let $\varphi \in \mathrm{Dom}(\mathcal{N}^{1/2})$ with $\mathbb{E}[\varphi] = 0$ and let Z on $(\mathcal{S}', \mathscr{B}(\mathcal{S}'), \mu)$ have the standard normal distribution. Suppose $\mathscr{H}$ is a separating class of compex-valued test functions defined on $\mathbb{R}$ such that for $h \in \mathscr{H}$, both $\mathbb{E}h(\varphi)$ and $\mathbb{E}h(Z)$ exist and f_h is a bounded function with bounded continuous derivative f_h'. Then we have*

$$
d_{\mathscr{H}}(\varphi, Z) \le \left\{ \sup_{h\in\mathscr{H}} |f_h'|_\infty \right\} \int_{\mathcal{S}'} \left| 1 - \int_{-\infty}^{\infty} \partial_t\,\mathcal{N}^{-1}\varphi(x) \cdot \partial_t\varphi(x)\, dt \right| \mu(dx).
$$

If $\varphi(x) = (x, \eta)$ for $\eta \in \mathcal{S}$ and $x \in \mathcal{S}'$, then φ has the normal distribution with mean 0 and variance $|\eta|_0^2$. It is easy to see that

$$
\begin{aligned}
&\left\{ \sup_{h\in\mathscr{H}} |f_h'|_\infty \right\} \int_{\mathcal{S}'} \left| 1 - \int_{-\infty}^{\infty} \partial_t\,\mathcal{N}^{-1}\varphi(x) \cdot \partial_t\varphi(x)\, dt \right| \mu(dx) \\
&= \left\{ \sup_{h\in\mathscr{H}} |f_h'|_\infty \right\} \cdot \left| 1 - |\eta|_0^2 \right|.
\end{aligned}
$$

So, if $\varphi \sim N(0,1)$, the upper bound in Theorem 6.1 is zero. This shows that the bound is tight. The quantity $\sup_{h\in\mathscr{H}} |f_h'|_\infty$ can be bounded by applying Proposition 2.2. The function f_h' is continuous if h is. For total variation distance, (2.7) allows h to be such that both the real and imaginary parts of it are continuous and bounded between 0 and 1.

Theorem 6.1 provides a general bound for the normal approximation for $\varphi \in \mathrm{Dom}(\mathcal{N}^{1/2})$. It can be applied to produce explicit bounds for special cases of φ as in the work of Nourdin and Peccati [31]. However, we will not pursue this application in this chapter.

Acknowledgement

This work was partially supported by Grant R-146-000-182-112 from the National University of Singapore and also by Grant 104-2115-M-390-002 from the Ministry of Science and Technology of Taiwan.

References

1. A. D. Barbour, Stein's method and Poisson process convergence, *J. Appl. Probab.* 25(A) (1988), 175-184.
2. A. D. Barbour, Stein's method for diffusion approximation, *Probab. Th. Rel. Fields* 84 (1990), 297-322.
3. A. D. Barbour, L. H. Y. Chen, *An Introduction to Stein's Method*, Lect. Notes Ser. Inst. Math. Sci. Natl. Univ. Singap., vol. 4, Singapore Univ. Press, World Scientific, Singapore, 2005.
4. Y. M. Berezansky, Yu. G. Kondratiev, *Spectral Methods in Infinite Dimensional Analysis*, (in Russian), Naukova Dumka, Kiev, 1988. English translation, Kluwer Academic Publishers, Dordrecht, 1995.
5. S. Chatterjee, Fluctuations of eigenvalues and second order Poincaré inequalities. *Probab. Th. Rel. Fields*, 143 (2009), 1-40.
6. L. H. Y. Chen, Poisson approximation for dependent trials, *Ann. Probab.* 3 (1975), 534-545.
7. L. H. Y. Chen, Stein meets Malliavin in normal approximation, *Acta Math. Vietnam.* 40 (2015), 205-230.
8. L. H. Y. Chen, L. Goldstein, Q. M. Shao, *Normal Approximation by Stein's Method*, Probability and Its Applications, Springer, 2011.
9. L. H. Y. Chen, G. Poly, Stein's method, Malliavin calculus, Dirichlet forms and the fourth moment theorem.*Festschrift Masatoshi Fukushima* (Z-Q Chen, N. Jacob, M. Takeda and T. Uemura, eds.), Interdisciplinary Mathematical Sciences, vol. 17, World Scientific, 107-130.
10. L. H. Y. Chen and Q. M. Shao, Normal approximation under local dependence, *Ann. Probab.* 32 (2004), 1985-2028.
11. P.-C. Chu, *Stein's Method, Malliavin Calculus, Lévy White Noise Analysis, and their Applications in Financial Mathematics*, PhD Thesis, School of Mathematical Sciences, University of Nottingham, 2015.
12. I. M. Gel'fand, N. Y. Vilenkin, *Generalized Functions*, vol. 4, Academic Press, 1964.
13. L. Gross, Potentional theory on Hilbert space, *J. Funct. Anal.* 1 (1967), 123-181.
14. T. Hida, *Analysis of Brownian Functionals*, Carleton Mathematical Lecture Notes 13, 1975.
15. T. Hida, H.-H. Kuo, J. Potthoff, L. Streit, *White Noise: An Infinite Dimensional Calculus*, Kluwer Academic Publishers, 1993.
16. T. Hida, Si Si, *Lectures on White Noise Functionals*, World Scientific Pub. Co., 2008.

17. I. Kubo, S. Takenaka, Calculus on Gaussian white noises I, *Proc. Japan Acad. Ser. A Math. Sci.* 56 (1980), 376-380; Calculus on Gaussian white noises II, *Proc. Japan Acad. Ser. A Math. Sci.* 56 (1980), 411-416; Calculus on Gaussian white noises III, *Proc. Japan Acad. Ser. A Math. Sci.* 57 (1981), 433-437; Calculus on Gaussian white noise IV, *Proc. Japan Acad. Ser. A Math. Sci.* 58 (1982), 186-189.

18. H.-H. Kuo, Integration by parts for abstract Wiener measures, *Duke Math. J.* 41 (1974), 373-379.

19. H.-H. Kuo, *Gaussian Measures in Banach Spaces*, Lect. Notes in Math., vol. 463, Springer-Verlag, Berlin/New York, 1975.

20. H.-H. Kuo, Y.-J. Lee, Integration by parts formula and Stein lemma on abstract Wiener space, *Communications on Stochastic Anal.* Vol.5 No.2 (2011), 405-418.

21. Y.-J. Lee, Sharp inequalities and regularity of heat semigroup on infinite dimensional space, *J. Funct. Anal.* 71 (1987), 69-87.

22. Y.-J. Lee, On the convergence of Wiener-Itô decomposition, *Bull. Inst. Math. Acad. Sinica* 17 (1989), 305-312.

23. Y.-J. Lee, Generalized functions on infinite dimensional spaces and its application to white noise calculus, *J. Funct. Anal.* 82 (1989), 429-464.

24. Y.-J. Lee, Analytic version of test functionals, Fourier transform and a characterization of measures in white noise calculus, *J. Funct. Anal.* 100 (1991), 359-380.

25. Y.-J. Lee, A characterization of generalized functions on infinite dimensional spaces and Bargmann-segal analytic functions, In *"Gaussian Random Field"*, The Third Nagoya Lévy semiar, edited by K. Itô and T. Hida, 272-284, World Scientific, 1991.

26. Y.-J. Lee, Integral representation of second quantization and its Application to white noise analysis, *J. Funct. Anal.* 133 (1995), 253-276.

27. Y.-J. Lee, H.-H. Shih, A characterization of generalized Lévy white noise functionals, *Quantum Information and Complexity*, Eds. T. Hida, K. Saitô and Si Si World Scientific, 2004, 321-339.

28. Y.-J. Lee, H.-H. Shih, Analysis of generalized Lévy white noise functionals, *J. Funct. Anal.* 211 (2004), 1-70.

29. P. Malliavin, Stochastic calculus of variations and hypoelliptic operators, *Proc. Int. Symp. on Stoch. Diff. Equations*, Kyoto 1976, Kinokuniya, 195-263, 1978.

30. I. Nourdin, Lectures on Gaussian approximations with Malliavin calculus, *Sém. Probab.* XLV (2013), Springer, 3-89.

31. I. Nourdin, G. Peccati, Stein's method on Wiener chaos, *Probab. Th. Rel. Fields* 145 (2009), 75-118.

32. D. Nualart and G. Peccati, Central limit theorems for sequences of multiple stochastic integrals, *Ann. Probab.* 33 (2005), 177-193.

33. J. Potthoff, L. Streit, A characterization of Hida distributions, *J. Funct. Anal.* 101 (1991), 212-229.

34. R. Schatten, *Norm Ideals of Completely Continuous Operators*, Springer-Verlag, Berlin/Göttingen/Heidelberg, 1960.

35. W. Schoutens, *Stochastic Processes and Orthogonal Polynomials*, Lecture Notes in Statistics, vol. 146, Springer-Verlag, Berlin/Heidelberg, 2000.
36. W. Schoutens, Orthogonal polynomials in Stein's method, *J. Math. Anal. Appl.* 253 (2001), 515-531.
37. H.-H. Shih, On Stein's method for infinite-dimensional Gaussian approximation in abstract Wiener spaces, *J. Funct Anal.* 261 (2011),1236-1283.
38. Si Si, *Introduction to Hida Distributions*, World Scientific Pub. Co., 2012.
39. C. Stein, A bound for the error in the normal approximation to the distribution of a sum of dependent random variables, in: *"Proceedings of the Sixth Berkeley Symposium on Mathematical Statistics and Probability,"* II: Probability theory, 583-602, Univ. California Press, Berkeley, 1972.
40. C. Stein, *Approximation Computation of Expectation*, IMS Lect. Notes Monogr. Ser. 7, Inst. Math. Statist., Hayward, CA, 1986.

SENSITIVE HOMOLOGY SEARCHING BASED ON MTRAP ALIGNMENT

Toshihide Hara* and Masanori Ohya†

Department of Information Sciences
Tokyo University of Science
2641 Yamazaki, Noda, Chiba, 278-8510, Japan
* hara@is.noda.tus.ac.jp
† ohya@rs.noda.tus.ac.jp

We develop a sensitive homology searching method by introducing the MTRAP alignment algorithm. Traditional method, such as BLAST, uses a measure determined on the assumption that there is no intersite correlation on the sequences. On the other hand, our new method takes the correlation of consecutive residues.

Contents

1. Introduction

Homology searching is one of the most important techniques in life science. The tools of the homology searching enable researchers to compare a query sequence, i.e. their target DNA/protein sequence, with a database of sequences. As the sequences in database are annotated with some information such as its function and its biochemical property, researchers estimate the identity of the target sequence by homology searching. BLAST [1, 2] is the most widely used tool of the homology searching developed by Altschul *et al.* in 1990. BLAST was much faster than others at the time and is still sufficiently fast. Although the speed is sufficient, the quality of results is not adequate in some cases.

We developed MTRAP sequence alignment algorithm in 2010 [3] which has been the most accurate aligner. Usual alignment methods in recent years use a measure empirically determined such as the measure defined by a combination of two quantities (1) and (2) below: (1) the sum of substitutions between two residue segments, (2) the sum of gap penalties in insertion/deletion region. Such a measure is determined on the assumption that there is no intersite correlation on the sequences. On the other hand, the MTRAP uses a unique approach that the correlations between two consective pairs of residues are considered. Our approach is based on the thermodynamic hypothesis: for a small globular protein, its three-dimensional structure is determined by the amino acid sequence of the protein. There may exist intersite correlations at least for two consecutive pairs of residues. Gonnet *et al.* reported this possibility [4]. Alignment method was improved by taking into account information of the intersite correlations. In this chapter, we introduce this approach into homology searching.

First of all, the MTRAP algorithm is shown in Section 2. After that we introduce a new homology searching method based on MTRAP in Section 3. After the review of benchmarking in Section 4, we conclude this chapter.

2. MTRAP

The MTRAP is an alignment algorithm by minimizing the value of a certain objective function based on the transition quantity [3, 5]. The algorithm is as follows.

Let Ω be the set of all amino acids, Ω^* be the Ω with the indel (gap) "*": $\Omega^* \equiv \Omega \cup \{*\}$. We call an element of Ω a residue and an element of Ω^* a symbol. In addition, let $\Gamma \equiv \Omega \times \Omega$ be the direct product of two Ωs and $\Gamma^* \equiv \Omega^* \times \Omega^*$.

Consider two arranged sequences, $A = a_1 a_2 \cdots a_n$ and $B = b_1 b_2 \cdots b_n$, both of length n, where $a_i, b_j \in \Omega^*$. Traditional alignment methods use a simple difference measure of A and B such as

$$d_{\text{sub}}(A, B) = \sum_i \tilde{s}(a_i, b_i),$$

where $\tilde{s}(a, b) \geq 0$ is a given difference of the symbol a and b. When the sequence A is equal to B the difference $d_{\text{sub}}(A, B)$ has a minimum value 0. On the other hand, the MTRAP uses the other difference of A and B such

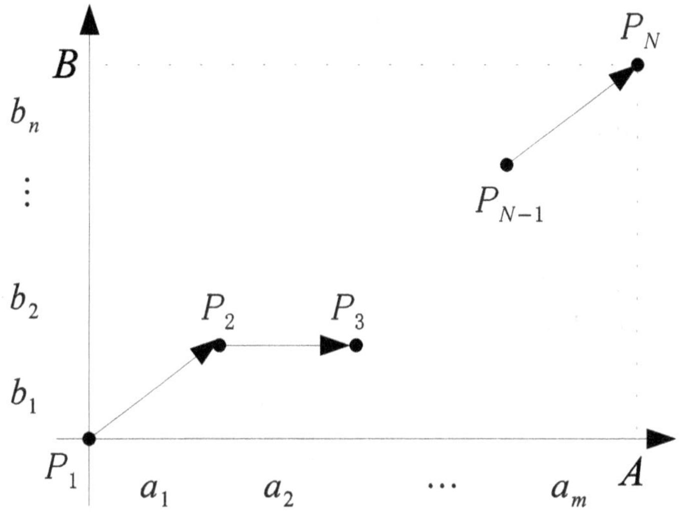

Fig. 1.   Example of the route for sequences A and B.

as

$$d_{\mathrm{MTRAP}}(A, B) = (1 - \varepsilon)\, d_{\mathrm{sub}}(A, B) + \varepsilon d_{\mathrm{trans}}(A, B),$$

$$d_{\mathrm{trans}}(A, B) = \sum_{i=1}^{n-1} \tilde{t}(u_i, u_{i+1}),$$

where $u_i = (a_i, b_i) \in \Gamma^*$ is the symbol pair of index i, $\tilde{t}(u_i, u_{i+1}) \geq 0$ is a given transition quantity between u_i and u_{i+1}, and $0 \leq \varepsilon \leq 1$ is a degree of mixture which is determined in advance w.r.t. the given $\tilde{s}$ and $\tilde{t}$.

The optimum alignment of given sequences A and B with the measure d_{MTRAP} is obtained as follows. Let $A = a_1 a_2 \cdots a_m$, $a_i \in \Omega$, $B = b_1 b_2 \cdots b_n$, $b_j \in \Omega$ be given two amino acid sequences. Take the lattice point $P_k = (i_k, j_k)$, $i = 1, \ldots, m$, $j = 1, \ldots, n$ as in Fig. 1. We call the sequence of the lattice points

$$\mathcal{R} = \{P_1, P_2, \ldots, P_N\}$$

a "route" with an initial point $P_1 = (0,0)$ and a final point $P_N = (m, n)$ if the following conditions are met:

$$i_{k-1} \leq i_k, \; j_{k-1} \leq j_k, \; P_{k-1} \neq P_k \text{ for any } k \, (= 2, 3, \ldots, N).$$

Let α_R, β_R be maps from a route $\mathcal{R} = \{P_1, P_2, \ldots, P_N\}$ to a set Ω^* such that

$$\alpha_R (P_k) = \begin{cases} a_{i_k} & i_k \neq i_{k-1} \ (k = 2, \ldots, N) \\ * & k = 1 \text{ or } i_k = i_{k-1} \ (k = 2, \ldots, N) \end{cases}$$

$$\beta_R (P_k) = \begin{cases} b_{j_k} & j_k \neq j_{k-1} \ (k = 2, \ldots, N) \\ * & k = 1 \text{ or } j_k = j_{k-1} \ (k = 2, \ldots, N) \end{cases}$$

and μ_R be a map from the route $\mathcal{R}$ to the set of all symbol pairs $\Gamma^* (\equiv \Omega^* \times \Omega^*)$ such that

$$\mu_R (P_k) = (\alpha_R (P_k), \beta_R (P_k)).$$

We call the following A^* and B^* the alignment of A and B by the route $\mathcal{R}$:

$$A^* : \alpha_R (P_1) \ \alpha_R (P_2) \ \cdots \ \alpha_R (P_N),$$
$$B^* : \beta_R (P_1) \ \beta_R (P_2) \ \cdots \ \beta_R (P_N).$$

Let $R(P)$ be the set of all routes with the final point P; that is,

$$R(P) = \{\{P_1, \ldots, P_k\}; \ P_k = P\}.$$

Let us fix the following notations for the following discussion: (1) $\Gamma^{*-} \equiv \Omega^* \times \Omega$, (2) $\Gamma^{-*} \equiv \Omega \times \Omega^*$, (3) $\Gamma^{g-} \equiv \{*\} \times \Omega$, (4) $\Gamma^{-g} \equiv \Omega \times \{*\}$, (5) W_{open} is a constant called gap "opening" cost; $0 \leq w_{\text{open}} \leq 1$, (6) w_{extend} is a constant called gap "extending" cost; $0 \leq w_{\text{extend}} \leq w_{\text{open}}$ and (7) ε is a weight; $0 \leq \varepsilon \leq 1$ (i.e., the mixture of usual difference d_{sub} and our new difference d_{trans}). The difference between A and B by a route $\mathcal{R}$ is given by

$$d(\mathcal{R}) = \sum_{k=2}^{N} \tilde{d}_s (P_{k-1}, P_k; \mathcal{R}),$$

where d_s is a function from $\Gamma^* \times \Gamma^*$ to $\mathbf{R}$ such that

$$\tilde{d}_s (P_{k-1}, P_k; \mathcal{R}) = \begin{cases} d_e (\mu_R (P_{k-1}), \mu_R (P_k)), & k \geq 3 \\ d_i (\mu_R (P_k)), & k = 2 \end{cases}$$

$$\tilde{d}_i (u) = \begin{cases} (1 - \varepsilon) \tilde{s} (u), & u \in \Gamma \\ (1 - \varepsilon) w_{\text{open}}, & u \notin \Gamma \end{cases}$$

$$
\tilde{d}_e\left(u_1, u_2\right) =
\begin{cases}
(1-\varepsilon)\,\tilde{s}\left(u_2\right) + \varepsilon \tilde{t}\left(u_1, u_2\right),\, u_1 \in \Gamma^*, u_2 \in \Gamma \\[2mm]
(1-\varepsilon)\,w_{\text{open}} + \varepsilon \tilde{t}\left(u_1, u_2\right),\,
\begin{cases}
u_1 \in \Gamma^{-*}, u_2 \in \Gamma^{g-} \\
u_1 \in \Gamma^{*-}, u_2 \in \Gamma^{-g}
\end{cases} \\[3mm]
(1-\varepsilon)\,w_{\text{extend}} + \varepsilon \tilde{t}\left(u_1, u_2\right),\,
\begin{cases}
u_1 \in \Gamma^{g-}, u_2 \in \Gamma^{g-} \\
u_1 \in \Gamma^{-g}, u_2 \in \Gamma^{-g}.
\end{cases}
\end{cases}
$$

The degree of difference between A and B with respect to a final point P can be defined as

$$
D\left(P\right) = \min_{R \in R(P)}\left\{d(\mathcal{R})\right\}.
$$

Hence the degree of difference between A and B is

$$
D_{AB} = D\left(P = (m, n)\right).
$$

We calculate D_{AB} by a dynamic programming technique as below. For a final point $P_k = (i, j)$ and a route $\mathcal{R} = \{P_1, \ldots, P_k\} \in R(P_k)$, we have

$$
P_k = (i, j) \text{ and } P_{k-1} = Q_1 \text{ or } Q_2 \text{ or } Q_3,
$$

where $Q_1 = (i-1, j)$, $Q_2 = (i-1, j-1)$ and $Q_3 = (i, j-1)$. Therefore

$$
R(P_k) = R_1(P_k) \cup R_2(P_k) \cup R_3(P_k)
$$

with

$$
R_l(P_k) = \{\{P_1, \ldots, Q_l, P_k\}\,;\, \{P_1, \ldots, Q_l\} \in R(Q_l)\}
$$

for $l = 1, 2, 3$. Thus we obtain

$$
\begin{aligned}
D\left(P = (m, n)\right) &= \min_{R \in R(P)}\left\{d(R)\right\} \\
&= \min_{l=1,2,3}\; \min_{R \in R_l(P)}\left\{d(R)\right\} \\
&= \min_{l=1,2,3}\; \min_{\substack{R_2 \in R(Q_l) \\ R_1 \in R_l(P)}}\left\{d(R_2) + d_s\left(Q_l, P; R_1\right)\right\} \qquad (2.1) \\
&= \min\left\{D_1(P), D_2(P), D_3(P)\right\},
\end{aligned}
$$

where

$$
D_l\left(P_k = (i, j)\right) \equiv \min_{\substack{R_2 \in R(Q_l) \\ R_1 \in R_l(P_k)}}\left\{d\left(R_2\right) + d_s\left(Q_l, P_k; R_1\right)\right\}
$$

for $l = 1, 2, 3$.

Each point Q_l has three points Q_l^1, Q_l^2, Q_l^3 which possibly go to Q_l one step after. These points are precisely written as

$$\begin{bmatrix} Q_1^1 & Q_1^2 & Q_1^3 \\ Q_2^1 & Q_2^2 & Q_2^3 \\ Q_3^1 & Q_3^2 & Q_3^3 \end{bmatrix}$$

$$= \begin{bmatrix} (i-2,j) & (i-2,j-1) & (i-1,j-1) \\ (i-2,j-1) & (i-2,j-2) & (i-1,j-2) \\ (i-1,j-1) & (i-1,j-2) & (i,j-1) \end{bmatrix}$$

when $Q_1 = (i-1,j)$, $Q_2 = (i-1,j-1)$, $Q_3 = (i,j-1)$.

The distances $D_l \left(P_k = (i,j) \right)$ can be obtained from one step before by the following recursion relations:

$$D_l \left(P_k = (i,j) \right) = \min_{\substack{R_1 \in R_l(P_k)}} \min_{\substack{p=1,2,3}} \min_{\substack{R_3 \in R(Q_l^p) \\ R_2 \in R_p(Q_l)}}$$

$$\left\{ d\left(R_3 \right) + d_s \left(Q_l^p, Q_l; R_2 \right) + d_s \left(Q_l, P_k; R_1 \right) \right\}$$

$$= \min_{\substack{R_1 \in R_l(P_k)}} \min_{\substack{p=1,2,3}} \left\{ D_p \left(Q_l \right) + d_s \left(Q_l, P_k; R_1 \right) \right\}$$

$$= \min_{\substack{p=1,2,3 \\ R_1 \in R_l(P_k)}} \left\{ D_p \left(Q_l \right) + d_s \left(Q_l, P_k; R_1 \right) \right\}$$

for $l = 1,2,3$. The values D_l of initial point and those of the edge points are assumed as

$$D_2 \left((0,0) \right) = 0,$$
$$D_l \left((0,0) \right) = \infty \text{ for } l = 1,3,$$
$$D_l \left((1,j) \right) = \infty \text{ for } l = 1,2,\ j = 1,\ldots,n,$$
$$D_l \left((i,1) \right) = \infty \text{ for } l = 2,3,\ i = 1,\ldots,m.$$

Moreover for other special cases, the recursive relation of the edge points satisfies

$$D_1 \left(P_k = (i,1) \right) = D_1 \left(P_{k-1} \right) + d_s \left(P_{k-1}, P_k; \mathcal{R} \right)$$

for $\mathcal{R} \in R_1 \left(P_k \right), i = 1,\ldots,m,$

$$D_3 \left(P_k = (1,j) \right) = D_3 \left(P_{k-1} \right) + d_s \left(P_{k-1}, P_k; \mathcal{R} \right)$$

for $\mathcal{R} \in R_3 \left(P_k \right), j = 1,\ldots,n.$

This calculation is completed in mn steps, and derives the optimum route. The optimum alignment can be generated as the pair of the two sequences A^* and B^* w.r.t. such the optimum route.

3. Homology Searching Based on MTRAP

We develop a homology searching algorithm based on MTRAP method. As an original MTRAP method is designed for global alignment, it can not be used as it is to homology search. The algorithm should be modified to determine similar regions between two sequences. One of the simplest ways is to introduce the Smith-Wateman's approach [6], i.e. to use a threshold for judging similar regions. We modified the algorithm as follows: (1) In place of $\tilde{s}(a_i, b_j)$ and $\tilde{t}(u_i, u_{i+1})$ of the previous section, we use $\tilde{s}(a_i, b_j) - \delta_s$ and $\tilde{t}(u_i, u_{i+1}) - \delta_t$ where δ_s, δ_t are the thresholds for judging similar regions. These values are empirically deduced by users. (2) The equation (2.1) is modified as $D(P = (m, n)) = \min\{0, D_1(P), D_2(P), D_3(P)\}$. Except these modifications, the dynamic programming technique described in the previous section can be used as it is. As the similar regions are expressed by minus value in routes of D_{AB}, we can obtain the regions by backtracing from the point with the value 0.

To compare the query sequence with the whole sequences of a certain database with the above modified algorithm, we can get the sequences similar to the partial of the query.

4. Performance Evaluation

For evaluating the quality of the proposal method, we use the following benchmark scheme. We use the well-treated dataset ASTRAL SCOP 2.03 (the version of with less than 40% identity to each other) [7], in which the family known sequences are stored. The whole sequences obtained from the dataset are used for both of the query for homology searching and the target database. we compare the precision values of our proposal with those of the BLAST. Precision is defined as $TP/(TP + FP)$, where TP means true positive (predicting that the sequences are homologous and it is true) and FP means false positive (predicting that the sequences are homologous but it is false). Fig. 2 shows that the precision values of our proposal and the BLAST w.r.t. each threshold value for the difference. The results indicate that the MTRAP could collect homologous sequences more precisely. Especially for the threshold > -4000, we can see the significant improvement. That is, our approach has a great effect on the searching for distantly related sequences.

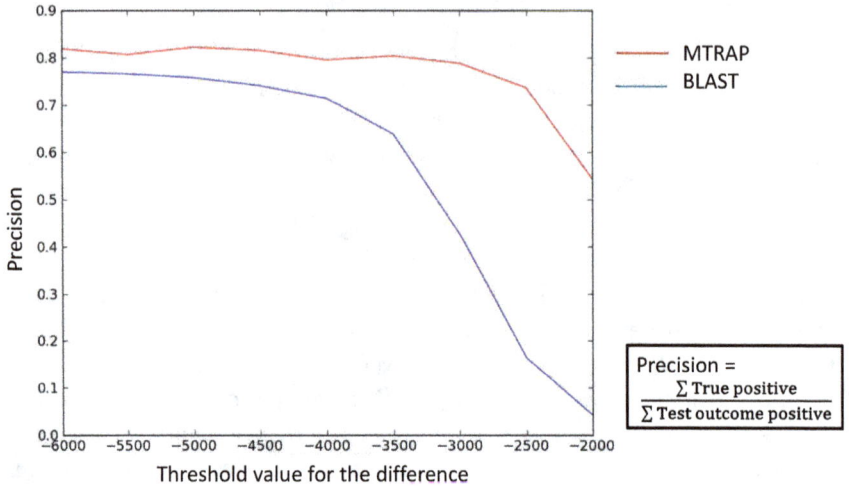

Fig. 2. The precisions of MTRAP and BLAST. The MTRAP (our proposal) is shown as red line and the BLAST is as blue line. Threshold values are scaled by multiplying 1000.

5. Conclusions

In our previous paper [3], we suggested some homework problems. Here we gave a solution for one of the problems, i.e. we introduced a high quality alignment called MTRAP into homology searching. To compare the precisions of our proposal with those of the BLAST, we indicated that our approach has a great effect on the searching for distantly related sequences. This result implies that the phylogenetic analysis among species of different phylesis becomes more reliable.

The other problems such as motif-findings will be discussed in our future paper.

References

1. S. F. Altschul, T. L. Madden, A. A. Schffer, J. Zhang, Z. Zhang, W. Miller, and D. J. Lipman, "Gapped BLAST and PSI-BLAST: a new generation of protein database search programs," Nucleic acids research, 25(17), pp. 3389-3402, 1997.
2. S. F. Altschul, W. Gish, W. Miller, E. W. Myers, and D. J. Lipman, "Basic local alignment search tool," J. Mol. Biol., 215(3), pp. 403-410, 1990.
3. T. Hara, K. Sato, and M. Ohya, "MTRAP: Pairwise sequence alignment algorithm by a new measure based on transition probability between two consecutive pairs of residues," BMC Bioinformatics, 11, 235, 2010.

4. G. Gonnet, and S. Benner, "Analysis of amino acid substitution during divergent evolution: the 400 by 400 dipeptide substitution matrix," Biochemical and Biophysical Research Communications, 199(2), pp. 489-496, 1994.

5. T. Hara, K. Sato, and M. Ohya, "Significant Improvement of Sequence Alignment can be Done by Considering Transition Probability between Two Consecutive Pairs of Residues," QP-PQ: Quantum Probability and White Noise Analysis (Quantum Bio-Informatics III), 26, pp. 443-452, 2010.

6. T. F. Smith, and M. S. Waterman, "Identification of common molecular subsequences," Journal of molecular biology, 147(1), pp. 195-197, 1981.

7. A. G. Murzin, S. E. Brenner, T. J. P. Hubbard, and C. Chothia, "SCOP: a structural classification of proteins database for the investigation of sequences and structures," J. Mol. Biol., 247, pp. 536-540, 1995.

SOME OF THE FUTURE DIRECTIONS OF WHITE NOISE THEORY

Takeyuki Hida

Professor Emeritus, Nagoya University, Japan
and Meijo University, Japan
thbty181@yahoo.co.jp

We shall discuss some of the future directions of white noise analysis. First, the space of generalized white noise functionals will be revisited from the viewpoint of an analog (not digital) stochastic calculus, which means the calculus of functions of continuously many independent idealized random variables.

Given a system of variables, it is natural to start with polynomials in those variables and to proceed to the space of generalized functions. In our case, those functions are to be generalized white noise functionals. Actual expressions of them are such that having the system of variables to be white noise, the time-derivatives $\dot{B} = \{\dot{B}(t), t \in R^1\}$, the functionals are expressed as $\varphi(\dot{B})$. Those functionals are naturally to be generalized functionals and they form a topological vector space.

We then proceed to the analysis of those functionals with the help of the harmonic analysis arising from the infinite dimensional rotation group.

In this line, we shall consider some of the new directions to be studied. Namely new noises are taken and the analysis of functionals of the noises will be studied. Further we shall just touch upon non-commutative calculus as a natural and new direction of the white noise theory.

AMS Subject Classification: 60H40 White Noise Theory

1. Introduction

The plan of this chapter:

1. The idea.

The basic idea of white noise theory is the *reductionism* for random complex systems. Namely, given a random complex phenomenon, we first find a system of independent random variables with the same information as the given phenomenon. Then, we form functions (usually functionals) of the independent random variables. The analysis of those functions leads us to the study of probabilistic properties of the system in question. Thus

Reduction $\rightarrow$ Functions $\rightarrow$ Analysis $\rightarrow$ Applications.

2. Our presentation is as follows:

i) Reductionism and noise.

ii) Functionals of a noise.

iii) Harmonic analysis. Useful transformation groups for noises.

iv) Towards non-commutative calculus.

v) Concluding remarks.

2. Reductionism and Noise

This section is devoted to a brief interpretation on the background of white noise analysis.

Following the reductionism, we first discuss on how to find a system of independent random variables. In most cases we take stochastic processes as random complex systems, so that we take Lévy's infinitesimal equation for a process $X(t)$. It is expressed in the form

$$\delta X(t) = \Phi(X(s), s \le t, Y(t), t, dt), \tag{1}$$

where $Y(t)$ stands for the (possibly infinitesimal) random variable, independent of $X(s), s \le t$ and contains the information that the $X(t)$ gains in the infinitesimal time interval $[t, t + dt)$. It is called the innovation of the $X(t)$.

This is, of course, a formal equation, but illustrates the idea to get the innovation of a process $X(t)$. The $\{Y(t)\}$ is exactly what we wish to have. See e.g. [8].

The innovation $Y(t)$ is, usually, an idealized random variable. We shall focus our attention to favorable cases, namely, satisfying stationarity in t. Then, it is an i.i.d. system. In addition, each $Y(t)$ is atomic in the usual sense. Such a system $\{Y(t)\}$ is called an idealized elemental random variables, abbr. i.e. r.v.'s, or simply called a "noise".

It is possible to classify the system of various noises. See e.g. [11]. A noise is parametrized in two ways: one is "time" parameter and the other is "space" parameter.

Making a long story short, there is a table:

Time parameter: Gaussian $\dot{B}(t)$, Poisson $\dot{P}(t)$

Space parameter: $P'(\lambda)$

Let $B(t)$ be a standard Brownian motion and let $P(t, \lambda)$ be a Poisson process with intensity λ, which may be viewed as a space parameter. The $P(\lambda) = P(1, \lambda)$ is obtained by fixing the time and by letting the intensity be a variable. We can even form a system $\{P(\lambda), \lambda \in (0, \infty)\}$ which is additive in λ, so that the derivative $P'(\lambda)$ can be defined as an idealized random variable.

To fix the idea we take, from what follows, in particular the white noise $\dot{B}(t)$.

3. Spaces of Generalized White Noise Functionals

1) Once variables, that is $\dot{B}(t)$'s are given, we come to discuss functions of them. We have so far discussed nonlinear functional of $\dot{B}(t)$'s as sums of multiple Wiener integrals. See e.g. [4] or [5]. This concept is fine, however we now change our viewpoint. We should regard multiple Wiener integrals as polynomials in $\dot{B}(t)$'s. For example, take a polynomial expressed in the form

$$\sum a_{(1,2,\cdots,n)} x_1^{p_1} x_2^{p_2} \cdots x_n^{p_n} \tag{2}$$

in discrete variables x_j's, regardless the x_j's are sure variables or independent random variables. Now we consider the continuous analog.

Start with the simplest case, that is the space of linear functions of the $\dot{B}(t)$'s. It is expressed in the form

$$\varphi(\dot{B}) = \int f(u) \dot{B}(u) du. \tag{3}$$

In the earlier notation it is to be

$$\int f(u) dB(u).$$

Following the reductionism, we are suggested to have single $\dot{B}(t)$, that is the case where $f(u) = \delta_t(u)$. For this purpose, we have a Gel'fand triple

$$\mathcal{H}_1^{(1)} \subset \mathcal{H}_1 \subset \mathcal{H}_1^{(-1)}$$

to define the generalized linear space $H_1^{(-1)}$ spanned by the $\dot{B}(t)$. This can be done by taking the isomorphism

$$\mathcal{H}_1^{(-1)} \cong K^{(-1)}(R^1),$$

where $K^{(-1)}(R^1)$ is the Sobolev space of order -1 over R^1. General notation: $K^{(m)}(R^n)$ is the Sobolev space of order m over R^n, with m non-negative or negative rational number.

Now each $\dot{B}(t)$ is a member of $H_1^{(-1)}$.

2) We then come to the next step to consider polynomials in the $\dot{B}(t)$'s of higher degree. Namely, we state our new viewpoint.

Since the number of the variables is continuously many (i.e. analog), a monomial looks like an integral in expression as in the linear case in the established theory. Our approach emphasizes that the integrals should be considered as polynomials in the $\dot{B}(t)$'s. Thus, the system of polynomials generates a vector space denoted by A; it forms a ring, in fact an integral domain.

We now wish, apart from the algebraic side, to proceed to the analysis. Note that there is a Gaussian measure μ behind, based on which the analysis should follow. Thus, it is natural to take a system of the Hermite polynomials in $\dot{B}(t)$'s that can form a base of the L^2-space defined by the generalized (idealized) Gaussian variables. Now we need rigorous justifications, as is seen below.

3) Since the basic variables are idealized random variables, it is necessary to give some interpretation to the nonlinear functions of them. Even to quadratic monomials, say $\dot{B}(t)^2$ we have to do something beyond a formal equality $\dot{B}(t)^2 = \frac{1}{dt}$ or equivalently $(dB(t))^2 = dt$. Note that the difference $(dB(t))^2 - dt$ is still random although infinitesimal, so we can not ignore the difference. We magnify it by multiplying $\frac{1}{(dt)^2}$ to have $(\dot{B}(t))^2 - \frac{1}{dt}$, that is the idealized Hermite polynomial $H_2(\dot{B}(t), \frac{1}{dt})$. Such an idea for modification is in agreement with the method of renormalization of polynomials in $\dot{B}(t)$'s.

4) We are going to generalize the classical story.

The collection of Hermite polynomials in the smeared variables $<\dot{B}, \xi_k>$'s span the Hilbert space (L^2) and admits the direct sum decomposition which is well-known as the Fock space:

$$(L^2) = \oplus \mathcal{H}_n,$$

where $\mathcal{H}_n$ is spanned by the Hermite polynomials in smeared variables. There is the isomorphism

$$\mathcal{H}_n \cong \sqrt{n!}\hat{L}^2(R^n),$$

where $\hat{L}^2$ means the symmetric L^2.

As has been shown in $[4, 5, 10]$ and others, we can define test functional space and generalized functional space of degree n, respectively:

$$\mathcal{H}_n^{(n)} \subset \mathcal{H}_n \subset \mathcal{H}_n^{(-n)},$$

where

$$\mathcal{H}_n^{(n)} \cong \sqrt{n!}\hat{K}^{(n+1)/2}(R^n).$$

The space $H_n^{(-n)}$ is a Hilbert space spanned by the Hermite polynomials (actually idealized Hermite polynomials with variance parameter $\frac{1}{dt}$) in $\dot{B}(t)$'s of degree n. We again use the notation $K^m(R^n)$ to express the Sobolev space and $\hat{K}$ for symmetric K.

The weighted sum $(L^2)^-$ of $H_n^{(-n)}$'s is the space of generalized white noise functionals. Again we emphasize the most significant understanding: $(L^2)^-$ is the collection of polynomials in $\dot{B}(t)$'s. The classical space H_n involves only multilinear functionals.

With these notes we will extend further developments of the white noise analysis as a future direction.

The space $(L^2)^+$ of test functionals is a suitable weighted sum of $H_n^{(n)}$ so as to be the dual space of $(L^2)^-$.

It is a good problem to establish characterizations of members in $(L^2)^-$, and those in $H_n^{(-n)}$. We hope, for computations of expectations, the ergodic property of the time shift operator acting on white noise space would be helpful for this purpose.

4. Renormalization and Further

The renormalization can be done rigorously by using the idealized Hermite polynomials with variance parameter $\frac{1}{dt}$. Let us remind

Theorem 4.1: *The renormalization is algebraically idempotent and it is defined so as to be a surjective onto the space* $(L^2)^-$ *of generalized white noise functionals.*

Here is a note. Denote by x a sample function of $\dot{B} = \{\dot{B}(t), t \in R^1\}$. The x runs through a space of generalized functions, say E^* the dual space of a nuclear space E which is dense in $L^2(R^1)$. The probability distribution μ of $\dot{B}$ is introduced on E^*. A functional of $\dot{B}$ can be represented by a measurable function $\varphi(x)$, $x \in E^*(\mu)$. Such functionals form a Hilbert space $(L^2)^-$ which is made to be isomorphic to $(L^2)^-$.

There is a transformation, called the S-transform. Let $\varphi(x)$ be in $(L^2)^-$. Then

$$(S\varphi)(\xi) = \exp[-\frac{1}{2}\|\xi\|^2] \int e^{<x,\xi>} \varphi(x) d\mu(x).$$

For example we have a representation of : $\dot{B}(t)^p$: in terms of $\xi(t)^p$. Generalizing this form a space of general white noise functionals, where we can carry on the calculus of white noise functionals in a quite natural way, by using the theory of functional analysis.

For example, it is not easy to define the $\dot{B}(t)$-derivative $\partial_t = \frac{\partial}{\partial \dot{B}(t)}$, because it is not a simple method to let $\dot{B}(t)$ change a little bit and to choose the topology when the limit of infinitesimal ratio is taken. To this end, having apply the S-transform, we apply the Fréchet derivative to functionals which is the image of the S-transform. The result is to be the S-transform of the $\dot{B}(t)$-derivative of the original $\dot{B}$-functional. One should be careful so as not to imitate the derivative on non-random functions.

We claim once again that such an approach comes from the fact of possible expressions of white noise functionals based on polynomials.

5. Some Useful Transformation Groups for the Noises

Having been motivated by the established theory of transformation groups, we can see, very often, the significant roles played by those groups in white noise theory. We shall choose, among others, four topics.

Before we come to the main topics, we need to make some general remarks. Noises that we are concerned are parameterized by continuous parameters, either the whole line R^1 or the positive half space. In any case, we are led to take the Affine group involving the shift and the dilation.

A noise has the probability distribution, so that the group of the measure-preserving transformations acts on the measure space that comes from the distribution.

Thus, we can think of two cases that lead us to, as it were, the harmonic analysis.

1) Gaussian case. If we make one point compactification so that we have the homeomorphism $\bar{R}^1 \cong S^1$, where we are given projective transformations, or conformal group. We may proceed to various one-parameter groups coming from the automorphisms of the parameter space.

Infinite dimensional rotation group $O(E)$.

After H. Yoshizawa, we define the rotation g of a nuclear space E.

Definition 5.1: A continuous linear transformation g acting on E is called a rotation of E, if it is an orthogonal transformation:

$$\|g\xi\| = \|\xi\|, \ \xi \in E,$$

where $\| \cdot \|$ is the $L^2(R^1)$-norm.

The collection of rotations of E forms a group. We denote the group by $O(E)$.

Definition 5.2: The group $O(E)$ is called the rotation group of E.

We are naturally given the adjoint group $O^*(E^*)$, which can characterize the white noise measure μ, which is the probability distribution of the $\{\dot{B}(t)\}$ that is introduced on E^*. This is the reason why the group $O(E)$ is very important in the study of white noise analysis.

Most of significant members of the group $O(E)$ are divided into two classes. One is the class I, which is determined as follows. Take a complete orthonormal system $\{\xi_n\}$, each member of which is in E. If $g \in O(E)$ is defined by a linear isometric (that is orthogonal) transformation defined by the $\{\xi_n\}$, then g is in class I. A member of class I is, in fact, digital.

While, a member g of class II is a transformation that comes from the automorphisms of the parameter space R^1. More explicitly

$$(g\xi)(t) = \xi(f(t))\sqrt{|f'(t)|},$$

where f is a smooth function that defines a surjection of R^1. A one-parameter subgroup of class II members is called a "whisker". Whiskers, often a group of them can serve essential roles in white noise analysis. Examples of such groups are those isomorphic to $PGL(2)$, $SO(d.1)$, $Aff(R^1)$ and so on. They are naturally introduced from our new viewpoint.

It is worth to be noted for other groups related to the non Gaussian case.

2) For a Poisson noise.

Beside the time shift, we can find a simple group that describes the duality between the time t and the intensity λ.

i) The Affine group $Aff(R^1)$ for a new noise of Poisson type.

Here we should like to emphasize the significance of this group.

Consider a noise depending on the space parameter, say λ associated with Poisson type noise as the intensity.

Observe how the group acts on distributions. Since the distribution of Poisson type may be viewed as a convergent sequence of positive numbers, we shall take the generating function that determines the sequence. Up to constant factor, it plays the key role of the representation of the Affine group. We are thus given an interesting problem to clarify the analytic role of the generating function in connection with the affine group.

The result is due to [11], so the details are skipped here.

ii) Compound noise.

A space noise has rather simple invariant properties under transformation groups, however, compound (space) noise enjoys interesting characters. For instance, if the intensity measure parametrized by scale parameter is formed so as to satisfy the dilation invariant, then arises a stable distribution. Such a formulation suggests us to find the so-called the underlying process, e.g. when a long tail distribution is observed.

6. Towards Non-Commutative White Noise Analysis

There are many directions that lead us to non-commutative white noise analysis which will be one of the important future research areas. We wish to find a road to this direction in a systematic manner. There are several examples of this direction, however we shall show a typical approach, that is the Feynman's path integral where white noise analysis plays a dominant role.

As for non-commutative analysis which has connection with white noise analysis, it is natural to be back to the famous literature [1]. By taking this spirit in mind we now discuss path integrals.

What we shall show in what follows is mainly in line with the W. Bock's approach, so that we wish to refer to [2] and avoid to explain details. It is interesting to note that the path integrals in this line are related to the original idea in the paper of [12], where we take the Brownian bridge to express possible trajectories around the classical path. This idea can well be realized by using the delta-function to express the effect of pinning the possible trajectories.

In what follows we shall just show that the white noise theory plays dominant roles. The main part for non-commutativity appears in the Hamiltonian path integrals, however we shall state briefly the Lagrangian path integral that we have established before. It is, of course well-known that there are relations between Lagrangian and Hamiltonian from mechanics. We wish to know any good probabilistic relations between the two kinds through the path integrals.

In the Hamiltonian dynamics, different from Lagrangian dynamics, the variables position (configuration) x and momentum p are independent variables, so that we should not understand as $p = m\frac{dq}{dt}$.

The relationship between x and q are connected by $dx \wedge dp$, where non-commutativity appears.

Before we come to the Hamiltonian path integral, we shall have a quick review of the earlier approach to the path integral using Lagrangian. See [12].

We set

$$S(t_0, t_1) = \int_{t_0}^{t_1} L(t)dt \tag{4}$$

and set

$$\exp\left[\frac{i}{\hbar} \int_{t_0}^{t_1} L(t)dt\right] = \exp\left[\frac{i}{\hbar} S(t_0, t_1)\right] = B(t_0, t_1).$$

Then, we have, for $0 < t_1 < t_2 < \cdots < t_n < t$,

$$B(0, t) = B(0, t_1) \cdot B(t_1, t_2) \cdots B(t_n, t).$$

(We may compare with the Markov property.)

Theorem 6.1: *The quantum mechanical propagator $G(0, t; y_1, y_2)$ is given by the following average*

$$G(0, t; y_1, y_2) = \left\langle Ne^{\frac{i}{\hbar} \int_0^t L(y, \dot{y})ds + \frac{1}{2} \int_0^t \dot{B}(s)^2 ds} \delta_o(y(t) - y_2) \right\rangle, \tag{5}$$

where N is the amount of multiplicative renormalization. The average $\langle \ \rangle$ is understood to be the integral with respect to the white noise measure μ.

Now we are in a position to discuss the Hamiltonian path integral. We follow the Klauder-Grothaus-Bock's line.

The Hamiltonian $H(x, p, t)$ is given by

$$H(x, p, t) = \frac{1}{2m} p^2 + V(x, p, t).$$

Hamiltonian action $S(x, p, t)$ is of the form

$$S(x, p, t) = \int_0^t p(\tau)\dot{x}(\tau) - H(x(\tau), p(\tau), \tau)d\tau.$$

First we give the configuration (coordinate space) path integral, then come to that on the momentum space.

1. Configuration space.

A possible trajectory starting from x_0 is denoted by

$$x(\tau) = x_0 + \sqrt{\hbar/m}B(\tau), 0 \leq \tau \leq t. \tag{6}$$

Momentum p is taken to be

$$p(\tau) = \sqrt{\hbar m}\omega(\tau), 0 \leq \tau \leq t,$$

where ω is another white noise.

Then, the Feynman integrand I_c is given by

$$I_c = N \exp \left[\frac{i}{\hbar} \int_0^t p(\tau)\dot{x}(\tau) - \frac{p(\tau)^2}{2m} d\tau + \frac{1}{2} \int_0^t \dot{x}(\tau)^2 + p(\tau)^2 d\tau \right]$$

$$\cdot \exp \left[-\frac{i}{\hbar} \int_0^t V(x(\tau), p(\tau), \tau) d\tau \right] \delta(x(t) - y),$$

where N is the (multiplicative) renormalizing constant.

2. Hamiltonian path integrals on the momentum space.

The variable $p(\tau)$ has a fluctuation simply as a Brownian motion

$$p(\tau) = p_0 + \frac{\sqrt{\hbar m}}{t} B(\tau), 0 \leq \tau \leq t.$$

The same for the space variable $x(\tau)$:

$$x(\tau) = \sqrt{\hbar/mt}\omega(\tau), 0 \leq \tau \leq t.$$

The Feynman integrand I_m is given by

$$I_m = N \exp \left[\frac{i}{\hbar} \int_0^t (-x(\tau)\dot{p}(\tau) - \frac{p(\tau)^2}{2m}) d\tau + \frac{1}{2} \int_0^t (\omega(\tau)^2 + B(\tau)^2) d\tau \right]$$

$$\cdot \exp \left[-\frac{i}{\hbar} \int_0^t V(x(\tau), p(\tau), \tau) d\tau \right] \delta(p(t) - p').$$

Take the expectation to obtain the quantum mechanical propagator.

With these actual computation we can come to non-commutative analysis.

7. Concluding Remarks

I. Decomposition of Lévy processes; revisited.

It may be discussed in connection with a representation of the Affine group by focusing our attention to the compound Poisson part of a Lévy process. See [7].

II. Passage from digital to analogue to be continuously infinite.

The transition will produce profound, sometimes in implicit figures, so that we need deep considerations.

Polynomials in continuously many variables have been discussed, and further we should consider differential operators similarly.

A quotation from Hermann Weyl [13].

"Mathematics is the science of the infinite."

III. The number operator $\Delta_\infty = \int \partial_t^* \partial_t dt$ characterizes the subspace $H_n^{(-n)}$ involving homogeneous polynomials of degree n.

IV. A characterization of $H_n^{(-n)}$-functionals can be given by the number operator N (or the Laplace-Beltrami operator Δ_∞ by putting $-$ in front, i.e. $\Delta_\infty = -N$).

Proposition 7.1: *On the space $(L^2)^-$ the pair $\{N, n\}$ forms an eigensystem of the number operator: for $\varphi(\dot{B})$ in $H_n^{(-n)}$ we have*

$$\mathcal{N}\varphi = n\varphi.$$

V. Space of the generalized white noise functionals. In this chapter we have introduced the space $(L^2)^-$, which is fitting to explain other topics in this chapter. There is, however, another beautiful method to introduce the space of generalized functionals. That is

$$(\mathcal{S}) \subset (L^2) \subset (\mathcal{S})^*.$$

This method has other advantages to discuss white noise analysis. We shall discuss in another opportunity.

References

1. V. I. Arnold, Mathematical methods of classical mechanics. 2nd ed. Springer, 1978.
2. W. Bock, Hamiltonian path integrals in white noise analysis. Dissertation, Univ. Kaiserslautern. 2013.
3. T. Hida, Analysis of Brownian functionals. Carleton Univ. Math. Notes 13, 1975,
4. T. Hida, H.-H. Kuo, J. Potthoff and L. Streit, White noise. An Infinite-dimensional calculus. Kluwer Academic Pub. 1993.
5. T. Hida and Si Si, Lectures on white noise functionals. World Sci. Pub. Co. 2008.
6. T. Hida, White noise (in Japanese). Maruzen Pub. Co., 2014. English translation by Hida and Si Si to appear.
7. P. Lévy, Sur les intégrales dont les éleménts sont des variables aléatoires indépendants. Ann. della R. Scuola Normale Superior di Pisa. Ser. II, III (1934) 337-366. Also three C.R. papers (1934).

8. P. Lévy, Random functions: General theory with with special reference to Laplacian random functions. Univ. of California Pub. in Statistics 1 (1953) 331-388.

9. J. Potthoff and L. Streit, A characterization of Hida distribution. J. Functional Analysis. 101. (1991), 212-229.

10. Si Si, Introduction to Hida distributions. World Scientific Pub. Co. 2012.

11. Si Si, Graded rings of homogeneous chaos generated by polynomials in noises depending on time and space parameters, respectively. Moscow Conf. (2013).

12. L. Streit and T. Hida, Generalized Brownian functionals and Feynman integral. Stochastic Processes and their Applications. 16 (1983), 55-69.

13. H. Weyl, Levels of infinity. Selected writings of infinity (1930). ed. P. Pesic. Dover Pub. 2012.

LOCAL STATISTICS FOR RANDOM SELFADJOINT OPERATORS

Peter D. Hislop

Department of Mathematics, University of Kentucky
Lexington, Kentucky 40506-0027, USA
hislop@ms.uky.edu

Maddaly Krishna

Institute for Mathematical Sciences, IV Cross Road, CIT Campus, Taramani
Chennai 600 113, Tamil Nadu, India
krishna@imsc.res.in

In this chapter we give a brief announcement of some of the results obtained recently in the eigenvalue statistics associated with some random selfadjoint operators, specifically the random Schrödinger and Anderson models.

Keywords: Anderson model; eigenvalue statistics.

1. Introduction

In the last few years local statistics of random Schrödinger operators and their discrete versions, known as the Anderson model, have steadily gained interest. Let $\{\omega_n : n \in \mathcal{I}\}$ be a collection of real valued independent, identically distributed random variables each distributed according to a probability measure μ. Then the models are the following:

$$H^\omega = \Delta + \sum_{n \in \mathcal{I}} \omega_n P_n \qquad (1)$$

where the operators (Δ, P_n) are defined as follows in the Schrödinger and Anderson cases respectively.

$$\Delta = -\sum_{j=1}^{d} \frac{\partial^2}{\partial x_j^2}, \quad P_n u(x) = \chi_{\Lambda(n)}(x) u(x), \; u \in L^2(\mathbb{R}^d)$$

with $\Lambda(n)$ a unit cube in $\mathbb{R}^d$ centered at n and $\mathcal{I} = \mathbb{Z}^d$;

$$(\Delta v)(n) = \sum_{|n-j|=1} v(j), \quad (P_n v)(m) = \sum_{|n-m|\leq k} A_{nm} v(n), \quad u \in \ell^2(\mathbb{Z}^d),$$

where A is a positive matrix of rank m_k and $\mathcal{I} = \mathbb{Z}^d/\Lambda_k$ with Λ_k a cube as in equation (2).

2. The Results

Given the above models, we are interested in the local statistics of the eigenvalues of these operators in the part of the spectrum that is pure point. It is known that when the disorder is small, which in the case when μ is absolutely continuous with bounded density ρ, means that the $\|\rho\|_\infty$ is very small compared to 1, there is pure point spectrum for the Anderson model [1], while there is pure point spectrum at the bottom of the spectrum in the Schrödinger case [7]. The problem is then to find out how the eigenvalues behave near a point in the point spectrum.

To study such an object, we need to look at finite dimensional approximants of these operators, since the point spectrum of these operators is dense. To this end we look at a cube

$$\Lambda_L = \begin{cases} \{x \in \mathbb{R}^d : |x| \leq L\}, & \mathbb{R}^d \text{ case}, \\ \{x \in \mathbb{R}^d : |x| \leq L\} \cap \mathbb{Z}^d, & \mathbb{Z}^d \text{ case}. \end{cases} \tag{2}$$

Then we look at the orthogonal projections

$$P_L = \begin{cases} \text{projection onto } \mathrm{L}^2(\Lambda_L), & \mathbb{R}^d \text{ case}, \\ \text{projection onto } \ell^2(\Lambda_L), & \mathbb{Z}^d \text{ case}. \end{cases}$$

We look at the operators associated with the orthogonal projections P_L and a point E in the point spectrum given by

$$H_L^\omega = P_L H^\omega P_L$$

which are selfadjoint matrices in the $\mathbb{Z}^d$ case and selfadjoint operators in the $\mathbb{R}^d$ case respectively. Let $E_A()$ denote the projection valued spectral measure associated with the selfadjoint operator A given by the spectral theorem and let $|\Lambda_L|$ denote the volume and the cardinality of Λ_L respectively in the continuous and discrete cases.

With this notation we look at the random measures

$$\xi_{L,E}^\omega(\cdot) = Tr\big(P_L E_{|\Lambda_L|(H_L^\omega - E)}(\cdot)\big). \tag{3}$$

These measures denote the way the eigenvalues of the approximants H_L^ω of the operators H^ω are located in the neighborhood of a point E in the point spectrum of H^ω. The scaling $|\Lambda_L|$ is to separate these eigenvalues sufficiently, as L increases, so that the local statistics stands out.

The spectral types are almost surely constant under the above conditions and let us denote Σ_p to be the almost sure point spectrum of the above model.

We say that E is in the region of complete exponential localization for the model (1) if there is an interval $[a,b]$, $\Sigma_p \cap [a,b] \neq \emptyset$ containing E such that for some $0 < s < 1$,

$$\sup_{Re(z) \in [a,b], Im(z) > 0} \mathbb{E}\big(|\langle \delta_n, (H_l^\omega - z)^{-1} \delta_m \rangle|^s\big) \leq C e^{-\gamma |n-m|}, \quad \text{for some } \gamma > 0,$$

(4)

in the discrete case and similarly for the continuous case where the argument in the expectation in equation (4) is replaced by

$$\|P_n (H_L^\omega - z)^{-1} P_m\|^s.$$

One of the main quantities that plays a role in the theory is the density of states, which is the density of the absolutely continuous measure

$$\mathcal{N}() = \mathbb{E}\big(Tr\big(Q E_{H^\omega}()\big)\big),$$

where $Q = P_0$ in the continuous case and is the orthogonal projection onto $\ell^2(\Lambda_k)$ in the discrete case.

This measure is known to be absolutely continuous and its density is denoted, n, is called the density of states.

The main result is then

Theorem 2.1: *Let H^ω be operators as in equation (1). Let $E \in \Sigma_p$ be in the region of complete exponential localization, such that the density of states $n(E) > 0$. Then, the limit points, in the sense of distributions, of the random measures $\xi_{L,E}^\omega$ are infinitely divisible point processes.*

In addition for each bounded borel set I:

- *In the discrete case, the limit points of $\xi_{L,E}^\omega(I)$ are compound Poisson random variables, with the corresponding Lévy measure supported on $\{1, 2, \ldots, m_k\}$.*
- *In the continuous case the limit points of the integer valued random variables $\xi_{L,E}^\omega(I)$ are compound Poisson distributed.*

The idea of the proof is to first show that the limit points of the sequence of random measures $\xi_{L,E}^\omega$ are infinitely divisible by comparing them to an asymptotically negligible array, which is then used through the Lévy-Khintichine formula to identify the Lévy measure of the limit points. Such an identification shows that the drift and the linear terms are absent in the exponent of the characteristic functions associated with the limit points.

The essential ingredients of the proofs are the Wegner estimate and the complete localization as stated here and an improved version of the Minami estimate which shows that the Lévy measure is finitely supported, in the set of integers, in the discrete case.

In the continuous model the earlier results [19] showed that the limit points are only infinitely divisible.

Initial work in showing Poisson statistics was done by Molchanov [18] in the one-dimensional case followed by Minami [17] in the d-dimensional discrete case, where he established the Minami estimate as a necessary step to show Poisson statistics (this bound in turn shows that the Lévy measure associated with the limiting process is supported on $\{1\}$.

Subsequently there are several works in this area [2, 12, 14, 5, 6, 20] showing Poisson statistics for various contexts.

Our result is the first to identify the possibility of having compound statistics and some examples are also given in [13], where the limiting point process is a compound Poisson process (meaning it is a Lévy process and the associated Lévy measure is supported on a single integer m_k).

Subsequent to our work Dolai-Krishna [11] showed Poisson statistics when the distributions of the random variables ω_n are not absolutely continuous, but are singular of various types.

Acknowledgement

MK wishes to thank Prof Si Si and Prof Watanabe for the invitation to the conference and the excellent hospitality provided by the Institute for Mathematical Sciences, National University of Singapore for local support which enabled him attend the conference. We also wish Prof Hida for a long and productive time ahead.

PDH was partially supported by NSF through grant DMS-1103104. MK was partially supported by IMSc Project 12-R&D-IMS-5.01-0106.

References

1. M. Aizenman, M. Stanislav, *Localization at large disorder and at extreme energies: an elementary derivation*, Commun. Math. Phys. **157(2)** (1993), 245–278.

2. M. Aizenman, S. Warzel, *The canopy graph and level statistics for random operators on trees*, Math. Phys. Anal. Geom. **9** (2006), no. 4, 291–333 (2007).

3. D. Applebaum, *Lévy processes and stochastic calculus*, Cambridge Studies in Advances Mathematics **116**, second edition, Cambridge: Cambridge University Press, 2009.

4. J. Bertoin, *Lévy processes*, Cambridge: Cambridge University Press, 1996.

5. J.-M. Combes, F. Germinet, A. Klein, *Generalized eigenvalue-counting estimates for the Anderson model*, J. Stat. Phys. **135** No. 2 (2009), 201–216.

6. J.-M. Combes, F. Germinet, A. Klein, *Poisson statistics for eigenvalues of continuum random Schrödinger operators*, Anal. PDE **3** (2010), no. 1, 49–80.

7. J.-M. Combes, P. D. Hislop, *Localization properties of continuous disordered systems in d-dimensions*, Mathematical Quantum Theory II, Schrödinger operators (Vancouer, BC, 1993). CRM Proc Lecture Notes **8**, Providence, RI: Am. Math. Soc., 1995, 213.

8. J.-M. Combes, P. D. Hislop, F. Klopp, *An optimal Wegner estimate and its application to the global continuity of the integrated density of states for random Schrödinger operators*, Duke Math. J. **140** (2007), 469–498.

9. J.-M. Combes, P. D. Hislop, F. Klopp, *New estimates on the spectral shift function for random Schrödinger operators*, Probability and mathematical physics, 85–95, CRM Proc. Lecture Notes, **42**, Amer. Math. Soc., Providence, RI, 2007.

10. S. De Bièvre, F. Germinet, *Dynamical localization for the random dimer model*, J. Stats. Phys. **98** (2000), 1135–1148.

11. D. R. Dolai, M. Krishna, *Level repulsion for a class of decaying random potentials*, Markov Process. Related Fields **21**, (2015), no. 3, part 1, 449–462.

12. F. Germinet, F. Klopp, *Spectral statistics for the random Schrödinger operators in the localized regime*, Jour. Eur. Math. Soc. (JEMS) **16** (2014), no. 9, 1967–2031. arXiv:1006.4427.

13. P. D. Hislop, M. Krishna, *Eigenvalue statistics for random Shroödinger operators with non-rank one perturbation*, Commun. Math. Phys. **340** (2015), no. 1, 125–143.

14. R. Killip, F. Nakano, *Eigenfunction statistics in the localized Anderson model*, Ann. Inst. Henri Poincare, **8(1)**, (2007) 27–36.

15. A. Klein, S. Molchanov, *Simplicity of eigenvalues in the Anderson model*, J. Stat. Phys. **122** (2006), no. 1, 95–99.

16. M. Krishna, P. Stollmann, *Direct integrals and spectral averaging*, J. Op. Theor. **69(1)**, (2013), 101–107.

17. N. Minami, *Local fluctuation of the spectrum of a multidimensional Anderson tight-binding model*, Commun. Math. Phys. **177** (1996), 709–725.

18. S. A. Molchanov, *The local structure of the spectrum of one-dimensional Schrödinger operator*, Commun. Math. Phys. **78** (1981), 429–446.

19. F. Nakano, *Infinite divisibility of random measures associated with some random Schrödinger operators*, Osaka J. Math. **46** (2009), 845–862.

20. M. Tautenhahn, I. Veselić, *Minami's estimate: Beyond rank one perturbation and monotonicity*, Ann. Henri Poincaré **15** (2014), 737–754.

MULTIPLE MARKOV PROPERTIES OF GAUSSIAN PROCESSES AND THEIR CONTROL

Win Win Htay

Department of Engineering Mathematics
Yangon Technological University
Yangon, Myanmar
htay22@googlemail.com

We shall first make a short survey on the multiple Markov properties of Gaussian processes, then come to the most general definition of these properties, where we use white noise theory, in particular, the recent results on generalized white noise functionals.

Having established the analytic properties of those multiple Markov Gaussian processes, we can observe some basic properties of those processes, then we shall come to some actual procedures to have the innovation as well as the best predictor of the future values based on the past observed data.

We also discuss the entropy loss which is one of the characteristics of multiple Markov Gaussian process expressing the rate of transmission of information.

Keywords: Gaussian process; Multiple Markov properties; white noise.

AMS Subject Classification 2000: 60H40

Contents

1. Introduction

The multiple Markov properties of Gaussian processes have been given by
T. Hida [1] in 1960, having been motivated by P. Lévy's research on Gaussian processes in 1960 at the 3rd Berkeley Symposium. Let $X(t), t \in T$
be a Gaussian process. The definition of its Markov property and multiple
Markov properties should be given in such a way that it expresses how
those random variables $X(t)$'s are dependent on each other. Since the $X(t)$
is Gaussian, their relationship can be described in terms of the correlation function $\Gamma(t, s)$ which exists for every pair (t, s). We may assume that
$E(X(t)) = 0$ identically, so that simply we consider

$$\Gamma(t, s) = E(X(t)X(s)).$$

Thus the multiple Markov properties shall be discussed basically in terms
of the covariance function.

Another background to be prepared comes from the main idea of white
noise analysis. It is the idea that is called the **reductionism**. Given a general (not necessarily to be restricted to a Gaussian system) random complex
system. To analyze such a system, we first form a system of **independent**
random variables that should contain the same information as the given
system. Then, the phenomena to be investigated should be expressed as
functions of the independent random variables that have been constructed.
We are thus ready to analyze the functions, and hence study the random
phenomena to be investigated.

Once the given system is restricted to be Gaussian, the main, actually
necessary, computations are linear, to our big advantage. Some more details
shall be discussed in the next section.

2. White Noise

Following the reductionism, we first try to find a system of independent
random variables that has the same information as the given random phenomena. In general, this problem is too hard to establish. One of the well-
known direction to approach is in line with the innovation theory. The idea
of this theory may be expressed in the form of the infinitesimal equation
proposed by P. Lévy. For a given stochastic process $X(t)$ we are interested
in the following formal equation

$$\delta X(t) = \Phi(X(s), s \le t, Y_t, t, dt), \tag{2.1}$$

where Y_t is the *innovation* which is an infinitesimal random variable which is independent of the $X(s)$, $s \leq t$ and contains the full new information that is gained by the process during the infinitesimal time interval $[t, t + dt)$. See P. Lévy [5]. This equation illustrates the structure of the process $X(t)$ from the viewpoint of the reductionism.

Although the formula (2.1) shows clear meaning, the practice to have the innovation is, in general, very difficult, and even so for the computations. There is, however, one exceptional case, where this idea can be realized: it is the Gaussian process (as well as fields or systems, etc.). We can appeal to the theory of the *canonical representations* of Gaussian processes, for which we can see how to form the innovation under our idea. We shall first state this fact in somewhat new style in the next section.

This chapter has another purpose. There may be attempts to give a definition of multiple Markov properties as a generalization of the simple Markov property. We take, among others, the definition of the multiple Markov property as is given in the next section. This definition can be naturally generalized to Gaussian random fields and even to generalized Gaussian processes under quite natural way of generalization. Further, we shall see the processes and fields specified by the definition are quite fitting for discussing their roles in information theory and for the forecasting problems of the systems in question as we shall see some details later.

3. Multiple Markov Properties, Revisited

We shall recall the definition of multiple Markov properties which should be a generalization of the simple Markov property. In the paper [1], Hida has given a definition of the N-ple Markov property of a Gaussian process. (The paper [1] is an old literature, however, we refer mainly for the historical interest.) At present we can rediscover the ideas behind the definition that was given more than fifty years ago. The ideas we can state in the following forms:

i) The definition has been given in connection with the aim of forecasting future values. There is involved the causality, namely the time propagation is always taken into account.

ii) Having been guided by the reductionism, we have a representation of a Gaussian process in terms of white noise. More precisely, the given Gaussian process $X(t)$ should be expressed as a functional of idealized elemental variables, more precisely as a function of independent identically distributed (i.i.d.) atomic random variables. Since we restrict our attention to Gaussian systems, we are suggested to use a white noise $\{\dot{B}(t), t \in R^1\}$, or almost equivalent, we may take Brownian motion $B(t), t \in R^1$. Then, we can discuss multiple Markov properties by using the representation of the Gaussian process $X(t)$ in question.

To fix the idea, we remind the definition of an N-ple Markov Gaussian process. We refer to [1]. Let $X(t)$ be a Gaussian process with $E(X(t)) = 0$ for every t. Let N be a positive integer.

Definition 3.1: If $X(t)$ satisfies the following two conditions for conditional expectations for any fixed t_0:

1. The conditional expectations $E(X(t_i)/\mathcal{B}_{t_0})$, $1 \leq i \leq N$ are linearly independent for any different t_i's,

2. While, $E(X(t_i)/\mathcal{B}_{t_0})$, $1 \leq i \leq N+1$ are linearly dependent for different t_i's,

then $X(t)$ is called N-ple Markov.

Being particularly related to the idea ii) listed above, we then come to a characterization of the N-ple Markov property of $X(t)$ and its analytic representation in terms of white noise. See also Hida [1] and Hida [2].

Theorem 3.2: *Suppose an N-ple Markov Gaussian process $X(t)$ has the canonical representation of the form*

$$X(t) = \int_0^t F(t, u)\dot{B}(u)du \qquad (3.1)$$

with an L^2-kernel $F(t, u)$ of Volterra type. Then, $F(t, u)$ is expressed in the form of Goursat kernel:

$$F(t, u) = \sum_1^N f_i(t)g_i(u).$$

For the later discussions, we remind the definition of a Goursat kernel. In the expression above, we state that it satisfies the following two conditions a) and b).

a) For any different t_j's, $det(f_i(t_j)) \neq 0$.

b) g_i's are linearly independent in $L^2([0, u])$ for any $u \geq 0$.

If $X(t), -\infty < t < \infty$ is a *stationary* Gaussian process, N-ple Markov property can be defined in the same idea. In this case $F(t, u)$ is expressed as $F(t - u)$ and the Fourier transform $\hat{F}$ of F is of the form

$$\hat{F}(\lambda) = \frac{Q(i\lambda)}{P(i\lambda)},$$

where P and Q are polynomials in $i\lambda$ and the degree of P is greater than that of Q.

There is another particular case. Let L_t be an Nth order ordinary differential operator, and let $X(t)$ satisfy the equation

$$L_t X(t) = \dot{B}(t), \ t \geq 0, \tag{3.2}$$

with the initial condition $X(0) = 0$. Then, we have a unique solution $X(t)$ which is an N-ple Markov Gaussian process. Tacitly, we assume some analytic conditions on the operator L_t. We often call the $X(t)$ *strictly N-ple Markov*.

With these background, although it seems old fashioned, we are highly motivated to study further directions to have further developments.

4. Generalized Gaussian Processes

Now comes the main part of this chapter.

What we have discussed so far can be extended to generalized Gaussian processes. There we shall see a best possible class of Gaussian processes where the multiple Markov properties can be introduced.

First, we give a definition of a generalized Gaussian process. We provide a nuclear space E which is dense in the Hilbert space $L^2(R^1)$ and a Probability space $(\Omega, P) = \Omega(P)$. Tacitly, the time parameter space is R^1.

Assume that

i) $X(\xi) = X(\xi, \omega), \xi \in E, \omega \in \Omega(P)$ is a Gaussian random variable and $\mathbf{X} = \{X(\xi), \xi \in E\}$ is a Gaussian system such that $E(X(\xi)) = 0$, and

ii) $X(\xi)$ is linear and strongly continuous in ξ in the space $L^2(\Omega, P)$.

Such a system $\mathbf{X}$ is called a *generalized Gaussian process*.

In order to discuss multiple Markov properties, we further introduce notations and assume necessary conditions as follows:

Let $\mathbf{B}_t(X)$ be the σ-field with respect to which all the $X(\xi)$'s with $supp(\xi) \subset (-\infty, t]$ are measurable. Set $\mathbf{B}(X) = \vee \mathbf{B}_t(X)$. Define the spaces $L_t(X) = L^2(\Omega, \mathbf{B}_t(X), P)$ and $L(X) = \vee L_t(X)$. The projection from $L(X)$ down to $L_t(X)$ is denoted by $E(t)$.

With these notations let us continue to have further assumptions.

iii) The space $L(X)$ is separable.

iv) $X(\xi)$ is purely non-deterministic, that is

$$\cap L_t(X) = \{0\}.$$

We are now ready to appeal to the Hellinger-Hahn theorem to have the direct sum:

$$L(X) = \bigoplus_n^m S_n,$$

where

$$S_n = \vee_t \{dE(t)Y_n, t \in R^1\},$$

that is, a cyclic subspace generated by some vector $Y_n \in L(X)$. Set $d\rho_n(t) = \|dE(t)Y_n\|^2$. Then the above S_n's are arranged in the decreasing order of $d\rho_n$, which guarantees the possibility of the multiplicity. It should not depend on the way of decomposition. Thus, the multiplicity is defined to be the maximum number m of subspaces S_n.

Our final assumptions are

v) $m = 1$, that is $X(\xi)$ has unit multiplicity and $d\rho_1$ is equivalent to the Lebesgue measure.

vi) $X(\xi)$ is continuous in ξ with respect to the Sobolev norm of order $-k$ with $k \geq 0$.

By the assumption v) we may assume that $X(\xi)$ is a continuous linear functional, or homogeneous polynomial in $\dot{B}(t)$. Namely, in terms of the white noise theory, $X(\xi)$ is a linear generalized white noise functional. Further, by iv), the kernel function of $X(\xi)$ is in the Sobolev space of order $-k$ over R^1. In other words, $X(\xi)$ is a liner homogeneous polynomial in the $\dot{B}(t)$'s, the coefficient is in the space $K^{(-k)}(R^1)$ depending linearly on ξ. We use the notation $K^p(R^n)$ to express the Sobolev space of order p over R^n.

We assume all the conditions mentioned above.

We are now ready to define the multiple Markov properties of $X(\xi)$.

Definition 4.1: A generalized Gaussian process $X(\xi)$ is N-ple Markov generalized Gaussian process if for any fixed t_0 and for any linearly independent ξ_i's with

$$supp(\xi_i) \subset [t_0, \infty)$$

the conditional expectations

1) $\{E(X(\xi_i)/\mathbf{B}_{t_0}(X)), 1 \leq i \leq N\}$ are linearly independent, and
2) $\{E(X(\xi_i)/\mathbf{B}_{t_0}(X)), 1 \leq i \leq N+1\}$ are linearly dependent.

To fix the idea, we assume that $X(\xi)$ is uniformly N-ple Markov, that is it is N-ple Markov for any time interval. Further we assume that $X(\xi)$ is continuous in $\xi \in K^N(R^1)$, where the notation $K^p(R^1)$ is the Sobolev space of order p (p can be any non-zero real number) over R^1. The last assumption implies that almost all sample function of X is in $K^{(-N)}(R^1)$.

We can now state a fundamental theorem basically due to Si Si [7] Section 7.5. with some modifications.

Theorem 4.2: *Let $X(\xi)$ be uniformly N-ple Markov. Then, there exist two systems of functions $\{f_i, 1 \leq i \leq N\}$ and $\{g_i, 1 \leq i \leq N\}$, respectively, such that each system involves N linearly independent functions in the symmetric $K^{(-N)}(R^1)$ and that $\det(< f_i, \xi_j >) \neq 0$ for any linearly independent ξ_j's.*

Further, we find a white noise $\dot{B}(t)$ such that

$$E(X(\xi)/\mathbf{B}_t(X)) = \sum_1^N < f_i, \xi > U_i^t,$$

where $\xi \in E$ and where $U_t^i = < g_i^t, \dot{B} >$, g_i^t being the restriction of g_i to $(-\infty, t]$.

Although the expression of the theorem is somewhat complicated, one can, however, easily see that this is in line with the idea of defining the N-ple Markov (ordinary) Gaussian process. It can be recognized that this result is best possible in generalization of multiple Markov properties so far as Gaussian is concerned. Thus, based on these observations, we can proceed to further investigations in the following section.

5. Forecasting and Controls of Multiple Markov Generalized Gaussian Processes

Given an N-ple Markov generalized Gaussian process $X(\xi)$ satisfying all the conditions in the last theorem. Take g_i's and apply a regularization of generalized function by using a test function η in such a way that: $\tilde{g}_i(u) = (g_i * \eta)(u)$. Thus, the enough analytic properties of functions involved in what follows are guaranteed, although the η is not written explicitly. The linearly independent property also holds for $\tilde{g}_i$'s, so that we can form a Frobenius formula for the $\tilde{g}_i$'s. There is defined a linear differential operator L_u^* of the Nth order such that

$$L_u^* \tilde{g}_i = 0, \ i = 1, 2, \ldots, N.$$

We can define a linear differential operator L_t which is the formal adjoint of L_u^*. Associated with this operator L_u^* is a fundamental system $\{\tilde{f}_i, 1 \leq i \leq N\}$ of solutions of

$$L_t \tilde{f}_i = 0, \ 1 \leq i \leq N.$$

Then, we can prove that there exist a matrix $A = A(t)$, which is in $GL(N, R^1)$ for every t, such that

$$(\tilde{f}_1, \ldots, \tilde{f}_N) = A(f_1, \ldots, f_N).$$

Replacing f_i's in the expression of $X(\xi)$ with $\tilde{f}_i$'s, we form a (generalized) Gaussian process $\tilde{X}(\xi)$. By using these facts we can prove

Theorem 5.1:

$$L_t \tilde{X}(\xi) = \dot{B}(\xi).$$

Proof comes from the fact that the kernel $\sum_1^N \tilde{f}_i(t) g_i(u)$ of Volterra type is the Riemann's function associated with the linear differential operator L_t.

By the actual computations the $\dot{B}(\xi)$, which is depending on the test function η, we can remove it.

Finally we note that we can find $\dot{B}(t)$ within the white noise theory. Again, noting the computations used above, we have

Corollary 5.2: *The $\dot{B}(t)$ is $\mathbf{B}_t(X)$-measurable.*

Thus, we have obtained the so-called *innovation* of X.

In line with the *causal calculus* where the time developments are always taken into account and where time order-preserving, we can freely discuss forecasting and control of X by using the annihilation operator $\partial_t = \frac{d}{d\dot{B}(t)}$ and the creation operator ∂_t^*.

Example 5.3: (Forecasting Prediction). Let t be the present time. We can form the best forecasting element of $X(\xi)$ base on the observed values up to present in the following element.

$$E(X(\xi)/\mathbf{B}_t(X)) = E(X(\xi)/\mathbf{B}_t(\dot{B})).$$

The right-hand side can be computed since we have actually obtained $\dot{B}(s), s \leq t$.

Acknowledgement

The author is grateful to "Institute for Mathematical Sciences of the National University of Singapore" for financial support and also to the organizers for the invitation to the "Workshop on Infinite Dimensional Analysis and Quantum Probability and Their Applications".

References

1. T. Hida, Canonical representations of Gaussian processes and their applications. Mem. Coll. of Sci. Univ. of Kyoto. Ser. A, 1960, 109-155.
2. T. Hida, Brownian motion. Springer Verlag. 1980.
3. T. Hida and Si Si, An innovation approach to random fields; Application of white noise theory. World Sci. Pub. Co. 2004.
4. T. Hida and Si Si, Lectures on White Noise Functionals. World Scientific Pub. Co. 2008.
5. P. Lévy, Random functions: General theory with special reference to Laplacian random functions. Univ. Calif. Pub. in Statistics. 1. 331-388. 1953.
6. P. Lévy, A special problem of Brownian motion, and a general theory of Gaussian random fields. Proc. 3rd Berkeley Symp. on Math. Statist. and Probability. vol. 2, 1956, 133-175.
7. Si Si, Introduction to Hida Distributions. World Scientific Pub. Co. 2011.
8. Si Si, Systems of idealized elemental random variables depending on space parameter. to appear.
9. Si Si and Win Win Htay, Entropy in subordination and filtering. Acta Applicandae Math. 63, 2000, 433-439.
10. N. Wiener, Nonlinear problems in random theory. The MIT Press, 1958.
11. Win Win Htay, Optimalities for random functions: Lee-Wiener's network and non-canonical representations of stationary Gaussian processes. Nagoya Math. J. 149, 1998, 9-17.
12. Win Win Htay, Multiple Markov property and Entropy of a Gaussian Process, JSPS Report, 2004, unpublished.
13. Win Win Htay, Multiple Markov properties for random fields. International Conference on Quantum Information and Complexity, Meijo Winter School, 2003, Nagoya, Japan.

14. Win Win Htay, Linear Process, Workshop on White Noise Analysis, Innovation, Complexity and Stochastic Mechanics, International Institute for Advanced Studies, 2003, Kyoto, Japan.

15. Win Win Htay, Note on Linear process, Proceedings of Quantum Information and Complexity, pp. 449-455, ed. T. Hida *et al.* World Scientific Pub. Co. 2004.

QUANTUM STOCHASTIC DIFFERENTIAL EQUATIONS ASSOCIATED WITH SQUARE OF ANNIHILATION AND CREATION PROCESSES

Un Cig Ji

Department of Mathematics
Institute for Industrial and Applied Mathematics
Chungbuk National University
Cheongju 28644, Korea
uncigji@chungbuk.ac.kr

Kalyan B. Sinha

Jawaharlal Nehru Centre for Advanced Scientific Research
Jakkur, Bangalore-64, India;
Department of Mathematics
Indian Institute of Science
Bangalore-12, India
kbs_jaya@yahoo.co.in

We develop the quantum stochastic calculus associated with the square of annihilation and creation processes. We study the quantum stochastic integral and quantum stochastic differential equation associated with the square of annihilation and creation processes. Under a sufficient condition the solution of the quantum stochastic differential equation becomes unitary and then the unitary solution gives a unitary dilation of an evolution system of which the Hamiltonian is depending on time.

1. Introduction

The quantum stochastic calculus as a quantum analogue of (classical) Itô calculus initiated by Hudson and Parthasarathy [9] has been developed extensively, in [2, 17, 18, 23, 24] with wide applications to mathematical physics. In the Hudson-Parthasarathy quantum stochastic calculus, the annihi-

---

Partly based on the lecures delivered in the IMS workshop on IDAQP, held in NUS, Singapore in 2014.

lation, creation and conservation processes play important roles as fundamental quantum stochastic processes, and then the unique solution of a certain quantum stochastic differential equation provides a stochastic dilation [3, 7, 18, 23, 24], called the Hudson-Parthasarathy dilation, of a uniformly continuous completely positive semigroup.

The main purpose of this chapter is to study a unitary dilation of an evolution system of which the Hamiltonian is depending on time. For our goal, we first develop the quantum stochastic integral associated with the square of annihilation and creation processes and then we prove the existence and uniqueness of solution of the quantum stochastic differential equation associated with the square of annihilation and creation processes. Then we establish a sufficient condition under which the solution of the quantum stochastic differential equation becomes unitary and then the unitary solution is a unitary dilation of an evolution system of which the Hamiltonian is depending on time. The paper [15] was published earlier than this chapter, but the paper [15] was written as generalizations of the main results of this chapter.

This chapter is organized as follows. In Section 2 we review basic notions for quantum stochastic processes on Fock space. In Section 3 we recall the Hudson-Parthasarathy quantum stochastic integral and study the quantum stochastic integral associated with the square of annihilation and creation processes. In Section 4 we establish the existence and uniqueness of solution of the quantum stochastic differential equation associated with the square of annihilation and creation processes. Finally, we study a sufficient condition under which the solution of the quantum stochastic differential equation becomes unitary.

Acknowledgements

The first author (UCJ) was supported by Basic Science Research Program through the NRF funded by the MEST (No. NRF-2013R1A1A2013712). The second author (KBS) gratefully acknowledges support from the J.N. Centre, Bangalore and the SERB-Distinguished Fellowship of DST, India.

2. Quantum Stochastic Processes

2.1. *Boson Fock space*

Let $\mathcal{H} = L^2(\mathbb{R}_+)$ be the Hilbert space of square integrable (complex-valued) functions of which the norm is denoted by $|\cdot|$ generated by the inner product $\langle \cdot | \cdot \rangle$ which is conjugate linear in the first variable and linear in the second variable. The *Boson Fock space* over $\mathcal{H}$ is defined by

$$\Gamma(\mathcal{H}) = \left\{ \phi = (f_n)_{n=0}^{\infty} \,\middle|\, f_n \in \mathcal{H}^{\widehat{\otimes}n}, \, \|\phi\|^2 = \sum_{n=0}^{\infty} n! \,|f_n|^2 < \infty \right\}$$

of which the inner product is denoted by $\langle\!\langle \cdot | \cdot \rangle\!\rangle$, where $\mathcal{H}^{\widehat{\otimes}n}$ is the nth symmetric tensor product of $\mathcal{H}$. For a dense subset $\mathcal{S}$ of $\mathcal{H}$, $\mathcal{E}(\mathcal{S})$ denotes the (dense) linear subspace of $\Gamma(\mathcal{H})$ generated by $\{\phi_\xi \in \Gamma(\mathcal{H}) \,|\, \xi \in \mathcal{S}\}$, where ϕ_ξ is an *exponential vector* or *coherent state* defined by $\phi_\xi = \left(1, \xi, \xi^{\otimes 2}/2!, \ldots, \xi^{\otimes n}/n!, \ldots\right)$.

For any $0 < s < t < \infty$, put

$$\mathcal{H}_{s]} = L^2([0,s], dt), \qquad \mathcal{H}_{[s,t]} = L^2([s,t], dt), \qquad \mathcal{H}_{[t} = L^2([t,\infty), dt),$$

and then, for each $0 < t_1 < t_2 < \cdots < t_n < \infty$, we have the following decomposition:

$$\mathcal{H} = \mathcal{H}_{t_1]} \oplus \mathcal{H}_{[t_1,t_2]} \oplus \cdots \oplus \mathcal{H}_{[t_{i-1},t_i]} \oplus \cdots \oplus \mathcal{H}_{[t_{n-1},t_n]} \oplus \mathcal{H}_{[t_n}.$$

Let $\mathfrak{h}$ be a fixed separable Hilbert space which is called the *initial Hilbert space*. Put

$$\mathfrak{H} = \mathfrak{h} \otimes \Gamma(\mathcal{H})$$

and, for any $0 < s < t < \infty$, put

$$\mathfrak{H}_{0]} = \mathfrak{h}, \quad \mathfrak{H}_{s]} = \mathfrak{h} \otimes \Gamma(\mathcal{H}_{s]}), \quad \mathfrak{H}_{[s,t]} = \Gamma(\mathcal{H}_{[s,t]}), \quad \mathfrak{H}_{[t} = \Gamma(\mathcal{H}_{[t}).$$

Then for each $0 < t_1 < t_2 < \cdots < t_n < \infty$, we have the following decomposition:

$$\mathfrak{H} = \mathfrak{H}_{t_1]} \otimes \mathfrak{H}_{[t_1,t_2]} \otimes \cdots \otimes \mathfrak{H}_{[t_{i-1},t_i]} \otimes \cdots \otimes \mathfrak{H}_{[t_{n-1},t_n]} \otimes \mathfrak{H}_{[t_n},$$

where the identity is determined by the following correspondence:

$$f \otimes \phi_\xi \quad \longleftrightarrow \quad f \otimes \phi_{\xi_{t_1]}} \otimes \phi_{\xi_{[t_1,t_2]}} \otimes \cdots \otimes \phi_{\xi_{[t_{i-1},t_i]}} \otimes \cdots \otimes \phi_{\xi_{[t_{n-1},t_n]}} \otimes \phi_{\xi_{[t_n}}$$

for all $f \in \mathfrak{h}$ and $\xi \in \mathcal{H}$. Here for each $\xi \in \mathcal{H}$,

$$\xi_{s]} = \mathbf{1}_{[0,s]}\xi, \quad \xi_{[s,t]} = \mathbf{1}_{[s,t]}\xi, \quad \xi_{[t} = \mathbf{1}_{[t,\infty)}\xi.$$

2.2. Fundamental quantum stochastic processes

For each given $t \geq 0$, we define linear operators A_t, $A_t^\dagger$ and Λ_t on $\mathcal{E}(\mathcal{H})$ by

$$A_t \phi_\xi := \left(\int_0^t \xi(s) ds \right) \phi_\xi,$$

$$A_t^\dagger \phi_\xi := \frac{d}{d\epsilon} \phi_{\xi + \epsilon \mathbf{1}_{[0,t]}} \Big|_{\epsilon=0},$$

$$\Lambda_t \phi_\xi := \frac{d}{d\epsilon} \phi_{e^{\epsilon \mathbf{1}_{[0,t]}} \xi} \Big|_{\epsilon=0} \tag{2.1}$$

for $\xi \in \mathcal{H}$, where for the definition of Λ_t, $\mathbf{1}_{[0,t]}$ is considered as the multiplication operator by $\mathbf{1}_{[0,t]}$. For an operator T in $\Gamma(\mathcal{H})$, $T^\dagger$ denotes the hermitian adjoint of T with respect to the inner product $\langle\!\langle \cdot | \cdot \rangle\!\rangle$. Then we can easily see that $(A_t)^\dagger = A_t^\dagger$ and $\Lambda_t^\dagger = \Lambda_t$. In fact, for any $\phi = (f_n) \in \Gamma(\mathcal{H})$ such that $f_n = 0$ except for finitely many n and $\xi \in \mathcal{H}$, we obtain that

$$\langle\!\langle \phi | A_t^\dagger \phi_\xi \rangle\!\rangle = \frac{d}{d\epsilon} \langle\!\langle \phi | \phi_{\xi + \epsilon \mathbf{1}_{[0,t]}} \rangle\!\rangle \Big|_{\epsilon=0}$$

$$= \sum_{n=0}^{\infty} \frac{d}{d\epsilon} \langle f_n | (\xi + \epsilon \mathbf{1}_{[0,t]})^{\otimes n} \rangle \Big|_{\epsilon=0}$$

$$= \sum_{n=1}^{\infty} n \langle f_n | \xi^{\otimes(n-1)} \widehat{\otimes} \mathbf{1}_{[0,t]} \rangle$$

$$= \sum_{n=0}^{\infty} (n+1) \langle f_{n+1} \widehat{\otimes}_1 \mathbf{1}_{[0,t]} | \xi^{\otimes n} \rangle,$$

where for $\zeta \in \mathcal{H}$, $f_{n+1} \widehat{\otimes}_1 \zeta \in \mathcal{H}^{\widehat{\otimes} n}$ such that

$$\langle f_{n+1} \widehat{\otimes}_1 \zeta | \xi^{\otimes n} \rangle = \langle f_{n+1} | \xi^{\otimes n} \widehat{\otimes} \zeta \rangle.$$

For each $\phi \in \mathbf{D}$, where

$$\mathbf{D} := \left\{ \psi = (f_n) \in \Gamma(\mathcal{H}) \ \Big| \ \sum_{n=0}^{\infty} (n+1)! \, (n+1) \, |f_{n+1}|^2 < \infty \right\},$$

we put

$$\widetilde{\phi} = \left((n+1) f_{n+1} \widehat{\otimes}_1 \mathbf{1}_{[0,t]} \right)_{n=0}^{\infty}$$

and then we have

$$\left\| \widetilde{\phi} \right\|^2 \leq t \sum_{n=0}^{\infty} n! (n+1)^2 |f_{n+1}|^2 = t \sum_{n=0}^{\infty} (n+1)! \, (n+1) |f_n|^2 < \infty$$

and so $\widetilde{\phi} \in \Gamma(\mathcal{H})$. Moreover, for any $\varphi \in \mathcal{E}(\mathcal{H})$, we have

$$\left| \left\langle\!\left\langle \phi \left| A_t^\dagger \varphi \right. \right\rangle\!\right\rangle \right| \leq \left\| \widetilde{\phi} \right\| \left\| \varphi \right\|.$$

Therefore, A_t can be extended as $(A_t^\dagger)^\dagger$ and its extension is denoted by the same symbol A_t. Then for any $\phi = (f_n) \in \mathbf{D}$, we have

$$A_t \phi = \widetilde{\phi} = \left((n+1) f_{n+1} \widehat{\otimes} \mathbf{1}_{[0,t]} \right)_{n=0}^\infty.$$

Moreover, we can easily see that A_t is a closed operator from $\mathbf{D}$ into $\Gamma(\mathcal{H})$ and so $A_t^\dagger$ is also a closed operator.

Lemma 2.1: ([9]) For each $s, t \geq 0$, we have

$$A_s A_t^\dagger = A_t^\dagger A_s + s \wedge t, \qquad A_s \Lambda_t = \Lambda_t A_s + A_{s \wedge t}.$$

Proof: The proof is straightforward by applying closedness of the annihilation operator and the definitions of A_t and Λ_t. $\qquad\square$

Remark 2.2: The first identity in Lemma 2.1 is called the *canonical commutation relation*.

Roughly speaking, a family $X = \{X_t \,|\, t \geq 0\}$ of (linear) operators in $\mathfrak{H}$ is called a *quantum stochastic process*. Then $A := \{A_t \,|\, t \geq 0\}$, $A^\dagger := \{A_t^\dagger \,|\, t \geq 0\}$ and $\Lambda := \{\Lambda_t \,|\, t \geq 0\}$ are called the *annihilation, creation and conservation processes*, respectively, and all together is called the *fundamental quantum stochastic processes*. For more detailed study, we refer to [9, 23].

2.3. *Adapted processes*

Let $\mathcal{D} \subset \mathfrak{h}$, $\mathcal{M} \subset \mathcal{H}$ be linear subspaces such that every $\xi \in \mathcal{M}$ is locally bounded with finite support, and $\mathbf{1}_{[s,t]}\xi \in \mathcal{M}$ for any $0 < s < t < \infty$ and $\xi \in \mathcal{M}$. Let $\mathfrak{M}$ be the linear subspace of $\mathfrak{H}$ generated by all vectors of the form $f \otimes \phi_\xi$ for $f \in \mathcal{D}$ and $\xi \in \mathcal{M}$.

A family $X = \{X_t \,|\, t \geq 0\}$ of operators in $\mathfrak{H}$ is called an *adapted (quantum stochastic) process* with respect to the triple $(\mathcal{D}, \mathcal{M})$ if

    (i) for any $t \geq 0$, $\mathrm{Dom}(X_t) \supset \mathfrak{M}$,
    (ii) for any $t \geq 0$, $f \in \mathcal{D}$ and $\xi \in \mathcal{M}$,

$$X_t(f \otimes \phi_{\xi_{t]}}) \in \mathfrak{H}_{t]} \quad \text{and} \quad X_t(f \otimes \phi_\xi) = \left(X_t(f \otimes \phi_{\xi_{t]}}) \right) \otimes \phi_{\xi_{[t}}.$$

If, in addition, for any $f \in \mathcal{D}$ and $\xi \in \mathcal{M}$, the map $\mathbb{R}_+ \ni t \mapsto X_t(f \otimes \phi_\xi) \in \mathfrak{H}$ is continuous, then $X = \{X_t \,|\, t \geq 0\}$ is said to be *regular* [23].

By direct computations from definitions of annihilation, creation and conservation processes, we can easily see that A, $A^\dagger$ and Λ are regular adapted with respect to $(\mathbb{C}, \mathcal{H})$ (see [9, 23]).

3. Quantum Stochastic Integrals

3.1. *HP-Quantum stochastic integrals*

In this section, we recall the quantum stochastic integral studied in [9] (see [23]). A quantum stochastic process X adapted with respect to $(\mathcal{D}, \mathcal{M})$ is said to be *simple* if there exists an increasing sequence $\{t_n \,|\, n = 0, 1, \ldots$ with $t_0 = 0$ and $t_n \to \infty$ as $n \to \infty\}$ such that

$$X(t) = \sum_{n=0}^{\infty} X(t_n) \mathbf{1}_{[t_n, t_{n+1})}(t). \tag{3.1}$$

If X is a simple adapted process satisfying (3.1) and M is one of the fundamental processes A, $A^\dagger$, Λ, then we define the linear operator:

$$\Xi(t) = \int_0^t X dM = \int_0^t X(s) dM(s),$$

which is called the *quantum stochastic integral* of X against with M, by putting
(i) $\mathrm{Dom}(\Xi(t)) = \mathcal{D} \otimes \mathcal{E}(\mathcal{M})$,
(ii) for any $f \in \mathcal{D}$ and $\xi \in \mathcal{M}$,

$$\Xi(t)(f \otimes \phi_\xi) = \begin{cases} (X(0)(f \otimes \phi_0)) \otimes M(t)\phi_\xi, & \text{if } 0 \leq t \leq t_1 \\ \Xi(t_n)(f \otimes \phi_\xi) + \big(X(t_n)(f \otimes \phi_{\xi_{t_n}})\big) \otimes (M(t) - M(t_n))\phi_{\xi_{[t_n}}, \\ \qquad \text{if } t_n < t \leq t_{n+1}, \quad n = 1, 2, \ldots, \end{cases}$$

where the right-hand side is defined inductively in n.

Note that the quantum stochastic integral $\Xi(t)$ of X is independent of the choice of partition to which X is simple, and $\{\Xi(t) | t \geq 0\}$ is a regular adapted process with respect to $(\mathcal{D}, \mathcal{M})$.

Then for the quantum stochastic integral

$$\Xi(t) = \int_0^t X(s) dM(s), \qquad t \geq 0,$$

and any $f \in \mathcal{D}$, $\xi \in \mathcal{M}$, by applying the fundamental lemma in [9] (see [23]) and the Gronwall's Inequality, we have

$$\| \Xi(t)(f \otimes \phi_\xi) \|^2 \leq e^{|\mu|([0,t])} \left(\int_0^t \| X(s)(f \otimes \phi_\xi) \|^2 \, d(|\nu| + |\mu|)(s) \right), \tag{3.2}$$

where measures μ and ν are given by the following tables:

M	A	$A^\dagger$	Λ
μ	$\overline{\mu_\xi}$	μ_ξ	$\mu_{\xi,\xi}$

M	A	$A^\dagger$	Λ
ν	0	μ_1 (Lebesgue measure)	$\mu_{\xi,\xi}$

$$(3.3)$$

with measures μ_ξ and $\mu_{\xi,\xi}$ given by

$$\mu_\xi([s,t]) = \langle \mathbf{1}_{[s,t]} \mid \xi \rangle, \qquad \mu_{\xi,\xi}([s,t]) = \langle \mathbf{1}_{[s,t]}\xi \mid \xi \rangle. \qquad (3.4)$$

Then the quantum stochastic integral is extended as followings. An adapted process X with respect to $(\mathcal{D},\mathcal{M})$ is said to be *measurable* if for any $f \in \mathcal{D}$ and $\xi \in \mathcal{M}$, the map $\mathbb{R}_+ \ni t \mapsto X(t)(f \otimes \phi_\xi)$ is Borel measurable. A $(\mathcal{D},\mathcal{M})$-adapted measurable process X is said to be *stochastically integrable* if there exists a sequence $\{X_n\}_{n=1}^\infty$ of simple $(\mathcal{D},\mathcal{M})$-adapted processes X_n such that

$$\lim_{n\to\infty} \int_0^t \| X_n(s)(f \otimes \phi_\xi) - X(s)(f \otimes \phi_\xi) \|^2 \, d|\tau| = 0 \qquad (3.5)$$

for any $t \geq 0$, $f \in \mathcal{D}$, $\xi \in \mathcal{M}$ and linear combinations τ of the Lebesgue measure and the measures μ_ξ and $\mu_{\xi,\xi}$ given as in (3.4).

The vector space of all such stochastically integrable processes is denoted by $L(\mathcal{D},\mathcal{M})$.

Proposition 3.1: ([23]) Let X be a $(\mathcal{D},\mathcal{M})$-adapted process satisfying the following conditions: for each $f \in \mathcal{D}$ and $\xi \in \mathcal{M}$,

(i) the map $\mathbb{R}_+ \ni t \mapsto X(t)(f \otimes \phi_\xi) \in \mathfrak{H}$ is left continuous,
(ii) $\sup_{0\leq s\leq t} \| X(s)(f \otimes \phi_\xi) \| < \infty$ for each $t \geq 0$.

Then $X \in L(\mathcal{D},\mathcal{M})$.

Let M be any one of the processes A, $A^\dagger$, Λ. Let $X \in L(\mathcal{D},\mathcal{M})$. Then there exists a sequence $\{X_n\}$ of simple processes X_n such that (3.5) holds and so by applying (3.2) we have

$$\left\| \int_0^t (X_m - X_n)dM(f \otimes \phi_\xi) \right\|^2$$
$$\leq e^{|\mu|([0,t])} \left(\int_0^t \| (X_m - X_n)(f \otimes \phi_\xi) \|^2 \, d(|\nu| + |\mu|)(s) \right) \qquad (3.6)$$

for all $f \in \mathcal{D}$ and $\xi \in \mathcal{M}$, where the measures μ and ν are given as in the tables in (3.3). Therefore, by applying (3.5) and (3.6), for any $f \in \mathcal{D}$ and

$\xi \in \mathcal{M}$, we can define

$$\left(\int_0^t X dM \right)(f \otimes \phi_\xi) := \lim_{n \to \infty} \left[\left(\int_0^t X_n dM \right)(f \otimes \phi_\xi) \right],$$

since the limit exists, and extend it linearly to the linear manifold $\mathcal{D} \otimes \mathcal{E}(\mathcal{M})$ of which the extension is denoted by $\int_0^t X dM$, which is called the *quantum stochastic integral* of X against with M. Note that if $X \in L(\mathcal{D}, \mathcal{M})$ and M is any one of the processes A, $A^\dagger$, Λ, then $\left\{ \int_0^t X dM \mid t \geq 0 \right\}$ is a regular $(\mathcal{D}, \mathcal{M})$-adapted process. Also, by Proposition 3.1, it is obvious that every regular $(\mathcal{D}, \mathcal{M})$-adapted process is stochastically integrable.

Proposition 3.2: ([9]) Let $j = 1, 2$, $X_j \in L(\mathcal{D}, \mathcal{M})$ and $M_j \in \{A, A^\dagger, \Lambda\}$. Then for the quantum stochastic integrals

$$\Xi_j(t) = \int_0^t X_j(s) dM_j(s), \qquad t \geq 0, \quad j = 1, 2,$$

and any $f, g \in \mathcal{D}$, $\xi, \eta \in \mathcal{M}$, we have

$$\langle\!\langle \Xi_1(t)(f \otimes \phi_\xi) | \Xi_2(t)(g \otimes \phi_\eta) \rangle\!\rangle = \sum_{k=1}^3 \int_0^t Y_k(s) d\mu_k(s), \qquad (3.7)$$

where

$$Y_1(s) = \langle\!\langle X_1(s)(f \otimes \phi_\xi) | \Xi_2(s)(g \otimes \phi_\eta) \rangle\!\rangle,$$
$$Y_2(s) = \langle\!\langle \Xi_1(s)(f \otimes \phi_\xi) | X_2(s)(g \otimes \phi_\eta) \rangle\!\rangle,$$
$$Y_3(s) = \langle\!\langle X_1(s)(f \otimes \phi_\xi) | X_2(s)(g \otimes \phi_\eta) \rangle\!\rangle,$$

and μ_1, μ_2 and μ_3 are given by the following tables:

M_1	A	$A^\dagger$	Λ
μ_1	$\overline{\mu_\xi}$	μ_η	$\mu_{\xi,\eta}$

M_2	A	$A^\dagger$	Λ
μ_2	μ_η	$\overline{\mu_\xi}$	$\mu_{\xi,\eta}$

(3.8)

$M_1 \backslash M_2$	A	$A^\dagger$	Λ
A	0	0	0
$A^\dagger$	0	μ_1 (Lebesgue measure)	μ_η
Λ	0	$\overline{\mu_\xi}$	$\mu_{\xi,\eta}$

(3.9)

respectively.

Let τ be a linear combination of the Lebesgue measure and the measures μ_ξ and $\mu_{\xi,\xi}$ given as in (3.4). Then for any simple $(\mathcal{D}, \mathcal{M})$-adapted process of the form given as in (3.1), define the integral $\int_0^t X d\tau$ of X with respect to τ by

$$\left(\int_0^t X d\tau \right)(f \otimes \phi_\xi) = \sum_{j=1}^n \tau([t_{j-1}, t_j]) X(t_{j-1})(f \otimes \phi_\xi) + \tau([t_n, t]) X(t_n)(f \otimes \phi_\xi),$$

if $t_n < t \le t_{n+1}$, for any $f \in \mathcal{D}$ and $\xi \in \mathcal{M}$. Then, by similar arguments, for $X \in L(\mathcal{D}, \mathcal{M})$, the integral $\int_0^t X d\tau$ is well-defined on $\mathcal{D} \otimes \mathcal{E}(\mathcal{M})$ and $\left\{ \int_0^t X d\tau \,\middle|\, t \ge 0 \right\}$ is a regular $(\mathcal{D}, \mathcal{M})$-adapted process. By similar arguments used in the proofs of Propositions 3.2, we obtain that for $\Xi_1(t) = \int_0^t X_1 d\tau$ and $\Xi_2(t) = \int_0^t X_2 dM$ with $X_j \in L(\mathcal{D}_j, \mathcal{M}_j)$ and $M \in \{A, A^\dagger, \Lambda\}$,

$$\langle\!\langle \Xi_1(t)(f \otimes \phi_\xi) | \Xi_2(t)(g \otimes \phi_\eta) \rangle\!\rangle = \int_0^t \langle\!\langle X_1(s)(f \otimes \phi_\xi) | \Xi_2(s)(g \otimes \phi_\eta) \rangle\!\rangle \, d\tau(s)$$

$$+ \int_0^t \langle\!\langle \Xi_1(s)(f \otimes \phi_\xi) | X_2(s)(g \otimes \phi_\eta) \rangle\!\rangle \, d\mu(s)$$

$$(3.10)$$

for all $f \in \mathcal{D}_1$, $\xi \in \mathcal{M}_1$, $g \in \mathcal{D}_2$ and $\eta \in \mathcal{M}_2$, where μ is given by μ_2 in (3.8).

Suppose $M_j \in \{A, A^\dagger, \Lambda\}$ $(j = 1, 2, \ldots, n)$, linear combinations τ_{j+n} $(j = 1, 2, \ldots, k)$ of the Lebesgue measure and measures μ_ξ and $\mu_{\xi,\xi}$ given as in (3.4), and $X_j \in L(\mathcal{D}, \mathcal{M})$ $(j = 1, 2, \ldots, n+k)$. Then

$$\Xi(t) = \sum_{j=1}^n \int_0^t X_j dM_j + \sum_{j=n+1}^{n+k} \int_0^t X_j d\tau_j, \qquad t \ge 0$$

is a regular $(\mathcal{D}, \mathcal{M})$-adapted process, and we write

$$d\Xi = \sum_{j=1}^n X_j dM_j + \sum_{j=n+1}^{n+k} X_j d\tau_j.$$

3.2. *Quantum stochastic integrals associated with square of annihilation and creation processes*

For each $t \ge 0$ and any $\xi, \eta \in \mathcal{H}$, we obtain that

$$\langle\!\langle A_t^2 \phi_\xi | \phi_\eta \rangle\!\rangle = \left(\int_0^t \xi(s) ds \right)^2 e^{\langle \xi | \eta \rangle} = 2 \left(\int_0^t \int_0^s \xi(u) \xi(s) du ds \right) e^{\langle \xi | \eta \rangle},$$

which implies the following quantum stochastic integral representation of the quantum martingale A_t^2:

$$\Delta_t := A_t^2 = 2 \int_0^t A_s dA_s. \qquad (3.11)$$

Similarly, we have

$$\Delta_t^\dagger := \left(A_t^\dagger\right)^2 = 2\int_0^t A_s^\dagger dA_s^\dagger. \tag{3.12}$$

For more study of the quantum stochastic integral representations of the quantum martingales, we refer to [11] and [13]. Now, we discuss the quantum stochastic integrals against with Δ_t and $\Delta_t^\dagger$.

Proposition 3.3: *Let X be a $(\mathcal{D}, \mathcal{M})$-adapted process satisfying the two conditions (i) and (ii) in Proposition 3.1. Then $X(\cdot) A_{(\cdot)} \in L(\mathcal{D}, \mathcal{M})$.*

Proof: For each $f \in \mathcal{D}$ and $\xi \in \mathcal{M}$, from the assumption (i), it is obvious that the map $t \longmapsto X(t)A_t(f \otimes \phi_\xi) = \left(\int_0^t \xi(u)du\right) X(t)(f \otimes \phi_\xi)$ is left continuous. Also, since every $\xi \in \mathcal{M}$ is locally bounded, for each $t \geq s \geq 0$, $f \in \mathcal{D}$ and $\xi \in \mathcal{M}$, we obtain that

$$\sup_{0 \leq s \leq t} \| X(s)A_s(f \otimes \phi_\xi) \| \leq \left(\int_0^t |\xi(u)|du\right) \sup_{0 \leq s \leq t} \| X(s)(f \otimes \phi_\xi) \|$$

is bounded. Therefore, by Proposition 3.1, $X(\cdot) A_{(\cdot)} \in L(\mathcal{D}, \mathcal{M})$. $\square$

From Proposition 3.3, for a $(\mathcal{D}, \mathcal{M})$-adapted process satisfying the conditions (i) and (ii) in Proposition 3.1, the integral $\int_0^t X(s)d\Delta_s$ is well-defined as following:

$$\int_0^t X(s)d\Delta_s := 2\int_0^t X(s)A_s dA_s.$$

For a $(\mathcal{D}, \mathcal{M})$-adapted process satisfying the following conditions: for each $f \in \mathcal{D}$ and $\xi \in \mathcal{M}$,

(i) the map $\mathbb{R}_+ \ni t \mapsto X(t)A_t^\dagger(f \otimes \phi_\xi) \in \mathfrak{H}$ is left continuous,
(ii) $\sup_{0 \leq s \leq t} \| X(s)A_s^\dagger(f \otimes \phi_\xi) \| < \infty$ for each $t \geq 0$.

Then by Proposition 3.1, $X(\cdot)A_{(\cdot)}^\dagger \in L(\mathcal{D}, \mathcal{M})$ and the integral $\int_0^t X(s)A_s^\dagger dA_s^\dagger$ is well-defined. Therefore, the integral $\int_0^t X(s)d\Delta_s^\dagger$ is defined by

$$\int_0^t X(s)d\Delta_s^\dagger := 2\int_0^t X(s)A_s^\dagger dA_s^\dagger.$$

3.3. *Quantum Itô formula*

In this section, we study a quantum counterpart of the classical Itô formula which is called a *quantum Itô formula*.

Theorem 3.4: ([9] **Quantum Itô Formula**) Suppose that $X_i \in L(\mathcal{D}, \mathcal{M})$ and $Y_j \in L(\mathcal{D}', \mathcal{M}')$ for $i, j = 1, 2, 3, 4$, and

$$d\Xi = \sum_{i=1}^{4} X_i dM_i = X_1 dA^\dagger + X_2 d\Lambda + X_3 dA + X_4 dt, \qquad (3.13)$$

$$d\Xi' = \sum_{j=1}^{4} Y_j dM_j = Y_1 dA^\dagger + Y_2 d\Lambda + Y_3 dA + Y_4 dt. \qquad (3.14)$$

Then for any $f \in \mathcal{D}$, $\xi \in \mathcal{M}$, and $g \in \mathcal{D}'$, $\eta \in \mathcal{M}'$, we obtain that

$$\langle\!\langle (\Xi(t) - \Xi(0)) (f \otimes \phi_\xi) | (\Xi'(t) - \Xi'(0)) (g \otimes \phi_\eta) \rangle\!\rangle$$

$$= \sum_{i=1}^{4} \int_0^t \langle\!\langle X_i(s)(f \otimes \phi_\xi) | \Xi'(s)(g \otimes \phi_\eta) \rangle\!\rangle \, \mu_i(s)$$

$$+ \sum_{j=1}^{4} \int_0^t \langle\!\langle \Xi(s)(f \otimes \phi_\xi) | Y_j(s)(g \otimes \phi_\eta) \rangle\!\rangle \, d\mu'_j(s)$$

$$+ \sum_{i=1}^{4} \sum_{j=1}^{4} \int_0^t \langle\!\langle X_i(s)(f \otimes \phi_\xi) | Y_j(s)(g \otimes \phi_\eta) \rangle\!\rangle \, d\mu_{ij}(s),$$

where μ_i, μ'_j and μ_{ij} are given by the following tables:

M_i	A	$A^\dagger$	Λ
μ_i	$\overline{\mu_\xi}$	μ_η	$\mu_{\xi,\eta}$

M_j	A	$A^\dagger$	Λ
μ'_j	μ_η	$\overline{\mu_\xi}$	$\mu_{\xi,\eta}$

(3.15)

$M_i \backslash M_j$	A	$A^\dagger$	Λ
A	0	0	0
$A^\dagger$	0	μ_1 (Lebesgue measure)	μ_η
Λ	0	$\overline{\mu_\xi}$	$\mu_{\xi,\eta}$

(3.16)

Let X_i, Y_j be processes in $L(\mathfrak{h}, \mathcal{H})$. Suppose X_i and X'_i are given as in (3.13) and (3.14), respectively, and $\Xi\Xi'$ is defined as a $(\mathcal{D}, \mathcal{M})$-adapted process. Furthermore, we assume that ΞY_j, $X_i \Xi'$ and $X_i Y_j$ are all defined as processes in $L(\mathcal{D}, \mathcal{M})$. Then as the differential version of Theorem 3.4 we have

$$d(\Xi\Xi') = \Xi(d\Xi') + (d\Xi)\Xi' + (d\Xi)(d\Xi'), \qquad (3.17)$$

where

$$\int_0^t \Xi d\Xi' = \sum_{j=1}^4 \int_0^t \Xi Y_j dM_j, \qquad \int_0^t (d\Xi)\Xi' = \sum_{i=1}^4 \int_0^t X_i \Xi' dM_i,$$

$$(d\Xi)(d\Xi') = \sum_{i,j=1}^4 X_i Y_j (dM_i)(dM_j),$$

and the Itô correction term $(dM_i)(dM_j)$ is calculated by the following formula:

$dM_i \backslash dM_j$	dA	$d\Lambda$	$dA^\dagger$	dt
dA	0	dA	dt	0
$d\Lambda$	0	$d\Lambda$	$dA^\dagger$	0
$dA^\dagger$	0	0	0	0
dt	0	0	0	0

Proposition 3.5: ([9, 23]) Suppose

$$d\Xi = \sum_{i=1}^4 X_i dM_i = X_1 dA^\dagger + X_2 d\Lambda + X_3 dA + X_4 dt,$$

where $X_i \in L(\mathcal{D}, \mathcal{M})$ for all $i = 1, 2, 3, 4$. Then for any $0 \le u < v \le T < \infty$, $f \in \mathcal{D}$ and $\xi \in \mathcal{M}$, we have

$$\| (\Xi(v) - \Xi(u))(f \otimes \phi_\xi) \|^2$$
$$\le 4 \left\{ \sum_{i=1}^4 e^{|\mu_i|([0,T])} \int_u^v \| X_i(s)(f \otimes \phi_\xi) \|^2 d(|\mu_i| + |\nu_i|)(s) \right\}, \quad (3.18)$$

where μ_i and ν_i are given by the following tables:

M_i	A	$A^\dagger$	Λ
μ_i	$\overline{\mu_\xi}$	μ_ξ	$\mu_{\xi,\xi}$

M_i	A	$A^\dagger$	Λ
ν_i	0	μ_1	$\mu_{\xi,\xi}$

4. Quantum Stochastic Differential Equations

Let $\mathrm{BC_s}([0,\infty), \mathcal{B}(\mathfrak{h}))$ be the space of all bounded strongly continuous maps $X : \mathbb{R}_+ \ni t \mapsto X(t) \in \mathcal{B}(\mathfrak{h})$. For each bounded strongly continuous map $X \in \mathrm{BC_s}([0,\infty), \mathcal{B}(\mathfrak{h}))$, we assign the adapted process $\{X(t) \otimes 1\}_{t \ge 0}$, where 1 is the identity operator in $\Gamma(\mathcal{H})$. For notational convenience, $X(t) \otimes 1$ is denoted by $X(t)$ again for each $t \ge 0$. Then $\{X(t)\}_{t \ge 0}$ is $(\mathfrak{h}, \mathcal{H})$-adapted.

Let $\{X_i\}_{i=1}^6$ be a family of maps in $BC_s([0,\infty), \mathcal{B}(\mathfrak{h}))$. Consider the following homogeneous quantum stochastic differential equation:

$$d\Xi = \Xi\left(X_1 d\Lambda^\dagger + X_2 dA^\dagger + X_3 d\Lambda + X_4 dA + X_5 d\Lambda + X_6 dt\right), \quad \Xi(0) = 1. \tag{4.1}$$

Then from (3.11) and (3.12), (4.1) is equivalent to

$$d\Xi = \Xi\left((2X_1 A^\dagger + X_2)dA^\dagger + X_3 d\Lambda + (X_4 + 2X_5 A)dA + X_6 dt\right), \quad \Xi(0) = 1, \tag{4.2}$$

which is is corresponding to the quantum stochastic integral equation: for all $t \geq 0$,

$$\Xi(s) = 1 + \sum_{k=1}^4 \int_0^t \Xi(s)Y_k(s)dM_k(s) \tag{4.3}$$

with

$$
\begin{array}{ll}
Y_1(s) = 2X_1(s)A_s^\dagger + X_2(s), & M_1(t) = A_t^\dagger, \\
Y_2(s) = X_3(s), & M_2(t) = \Lambda_t, \\
Y_3(s) = X_4(s) + 2X_5(s)A(s), & M_3(t) = A_t, \\
Y_4(s) = X_6(s), & M_4(t) = t.
\end{array} \tag{4.4}
$$

4.1. *Existence and uniqueness of solution*

For the given families $\{X_i\}_{i=1}^6$ of maps in $C([0,\infty), \mathcal{B}(\mathfrak{h}))$ the set of continuous maps, we have

$$K_t := \max_{1\leq i\leq 6} \sup_{0\leq s\leq t} \|X_i(s)\| < \infty \quad \text{for each} \quad t \geq 0. \tag{4.5}$$

Theorem 4.1: *There exists a unique regular $(\mathfrak{h}, \mathcal{H})$-adapted process $\Xi = \{\Xi(t) \,|\, t \geq 0\}$ satisfying (4.2) such that*

$$\sup_{0\leq t\leq T} \sup_{|f|\leq 1} \left\| \Xi(t)\left(\prod_{j=1}^n Y_{i_j}\right)(f \otimes \phi_\xi) \right\| < \infty, \quad \xi \in \mathcal{H}, \quad T \geq 0, \quad n \geq 1, \tag{4.6}$$

where $Y_{i_j} \in \{Y_1, Y_3\}$.

Proof: We prove this theorem for existence and uniqueness of solution of (4.2) by proving the existence and uniqueness of solution of (4.3) by using the standard Picard method with the estimates of Proposition 3.5. The proof is a simple modification of the proof of Proposition 26.1 in [23] and so we sketch the proof. Put

$$\Xi_0(t) = \Xi(0) = 1,$$

$$\Xi_n(t) = 1 + \sum_{i=1}^{4} \int_0^t \Xi_{n-1}(s) Y_i(s) dM_i(s) \tag{4.7}$$

for $n = 1, 2, \ldots$. In fact, for each n, Ξ_{n-1} is a regular $(\mathfrak{h}, \mathcal{H})$-adapted process, and $\Xi_{n-1} Y_i$ is in $L(\mathfrak{h}, \mathcal{H})$ and so Ξ_n is well-defined as a regular $(\mathfrak{h}, \mathcal{H})$-adapted process. Then by (4.7) we have

$$\Phi_n(t) = \int_0^t \Phi_{n-1}(s) \left(\sum_{i=1}^{4} Y_i(s) dM_i(s) \right), \quad \Phi_n(t) = \Xi_n(t) - \Xi_{n-1}(t)$$

and, by applying Proposition 3.5, for any fixed $T > 0$ and $0 \leq t \leq T$, we have

$$\| \Phi_n(t)(f \otimes \phi_\xi) \|^2 \leq 4 e^{\lambda_f([0,t])} \int_0^t \left(\sum_{i=1}^{4} \| \Phi_{n-1}(s)(Y_i(s) f \otimes \phi_\xi) \|^2 \right) \lambda_f(ds),$$

where λ_f is the measure obtained by summing up the Lebesgue measure and the $|\nu_i|$'s, the $|\mu_i|$'s that occur in the statement of Proposition 3.5. Then by induction in n, we obtain that

$$\| \Phi_n(t)(f \otimes \phi_\xi) \|^2 \leq 4^n e^{n \lambda_f([0,T])} \int_{0 < t_1 < t_2 < \cdots < t_n < t}$$

$$\times \sum_{\substack{1 \leq i_j \leq 4 \\ j = 1, 2, \cdots, n}} \| \Xi(0) S_{i_1}(t_1) S_{i_2}(t_2) \cdots S_{i_n}(t_n)(f \otimes \phi_\xi) \|^2$$

$$\times \lambda_f(dt_1) \lambda_f(dt_2) \cdots \lambda_f(dt_n),$$

where $S_i \in \{Y_1, \ldots, Y_4\}$ $(i = 1, 2, 3, 4)$. By using the non-atomicity of the measure λ_f we have

$$\| (\Xi_n(t) - \Xi_{n-1}(t))(f \otimes \phi_\xi) \|^2$$

$$\leq | f |^2 \| \phi_\xi \|^2 e^{2|\xi|^2} \frac{\left(\int_0^t |\xi(u)| du \vee 1 + T \vee 1 \right)^n 2^n (3K_T)^n \lambda_f([0,t])^n}{n!}, \tag{4.8}$$

where K_T is given as in (4.5), which implies that

$$\sum_{n=1}^{\infty} \sup_{0 \leq t \leq T} \| (\Xi_n(t) - \Xi_{n-1}(t))(f \otimes \phi_\xi) \| < \infty \tag{4.9}$$

for any $T \geq 0$, $f \in \mathfrak{h}$ and $\xi \in \mathcal{H}$. The inequality (4.9) ensures that the limit:

$$\lim_{n \to \infty} \Xi_n(t)(f \otimes \phi_\xi) = \Xi(0)(f \otimes \phi_\xi) + \sum_{n=1}^{\infty} (\Xi_n(t) - \Xi_{n-1}(t)) \, (f \otimes \phi_\xi) \quad (4.10)$$

exists uniformly on every bounded interval. Therefore, for each $t \geq 0$, define the operator $\Xi(t)$ in the domain $\mathfrak{h} \otimes \mathcal{E}(\mathcal{H})$ by

$$\Xi(t)(f \otimes \phi_\xi) = \lim_{n \to \infty} \Xi_n(t)(f \otimes \phi_\xi) \quad (4.11)$$

and extend it by linearity. In fact, from the uniform convergence of (4.10), the map $t \mapsto \Xi(t)(f \otimes \phi_\xi)$ is continuous for each $f \in \mathfrak{h}$ and $\xi \in \mathcal{H}$. Also, by direct computation applying the inequality (4.8), we can see that Ξ satisfies (4.6). Thus $\Xi = \{\Xi(t) \,|\, t \geq 0\}$ is a $(\mathfrak{h}, \mathcal{H})$-adapted regular process, and from (4.7), (4.11) and the definition of quantum stochastic integrals, Ξ satisfies (4.3). This proves the existence of a solution of (4.2).

Suppose that $\Xi' = \{\Xi'(t) \,|\, t \geq 0\}$, $\Xi'(0) = \Xi(0)$ is another solution of (4.2) satisfying (4.6) with $\Xi = \Xi'$. For each $t \geq 0$, put $\Upsilon(t) = \Xi(t) - \Xi'(t)$. Then we have

$$\Upsilon(t) = \sum_{i=1}^{4} \int_0^t \Upsilon(s) Y_i(s) dM_i(s),$$

and by applying Proposition 3.5 we have

$$\| \Upsilon(t)(f \otimes \phi_\xi) \|^2 \leq 4 e^{\lambda_f ([0,T])} \int_0^t \left(\sum_{i=1}^{4} \| \Upsilon(s)(Y_i(s) f \otimes \phi_\xi) \|^2 \right) d\lambda_f(ds) \quad (4.12)$$

for any $0 \leq t \leq T$, $f \in \mathfrak{h}$ and $\xi \in \mathcal{H}$ from which for each $T \geq t \geq 0$, by using (4.6) and the arguments employed in the proof of (4.8), we prove that there exists a constant $K(T, f, \xi) \geq 0$ depending on T, f and ξ such that

$$\| \Upsilon(t)(f \otimes \phi_\xi) \|^2 \leq \frac{K(T, f, \xi)^n}{n!}$$

for any $n \geq 1$. Letting $n \to \infty$ on the right-hand side, we conclude that $\Upsilon(t) = 0$ for $0 \leq t \leq T$. Since T is arbitrary, we have $\Upsilon(t)(f \otimes \phi_\xi) = 0$ for any $t \geq 0$, $f \in \mathfrak{h}$ and $\xi \in \mathcal{H}$, which implies that $\Upsilon = \Xi - \Xi' = 0$ on $\mathfrak{h} \otimes \mathcal{E}(\mathcal{H})$. $\qquad \square$

4.2. *Unitary solution with constant coefficients*

We study a special case of constant coefficients of (4.1) of which the solution $\Xi(t)$ is a unitary operator for each t. Consider the following equation:

$$d\Xi = \Xi\left(X_1 dA^\dagger + X_2 dA^\dagger + X_3 d\Lambda + X_4 dA + X_5 d\Delta + X_6 dt\right), \quad \Xi(0) = 1,$$
(4.13)

where X_j are constant adapted processes of the form $X_j(t) = X_j \otimes 1$ for $X_j \in \mathcal{B}(\mathfrak{h})$.

To find conditions under which the unique solution $\Xi(t)$ provided by Theorem 4.1 is unitary, we first assume that $\Xi(t)$ is an isometry on $\mathfrak{h} \otimes \mathcal{E}(\mathcal{H})$ for every $t \geq 0$, i.e.,

$$\langle\!\langle \Xi(t)(f \otimes \phi_\xi) | \Xi(t)(g \otimes \phi_\eta) \rangle\!\rangle = \langle\!\langle f \otimes \phi_\xi | g \otimes \phi_\eta \rangle\!\rangle \quad (4.14)$$

for all $t \geq 0$, $f, g \in \mathfrak{h}$, $\xi, \eta \in \mathcal{H}$, from which, by Theorem 3.4 and Proposition 3.2, we obtain that

$$\begin{aligned}
0 &= \langle\!\langle \Xi(t)(f \otimes \phi_\xi) | \Xi(t)(g \otimes \phi_\eta) \rangle\!\rangle - \langle\!\langle f \otimes \phi_\xi | g \otimes \phi_\eta \rangle\!\rangle \\
&= \langle\!\langle \Xi(t)(f \otimes \phi_\xi) | \Xi(t)(g \otimes \phi_\eta) \rangle\!\rangle - \langle\!\langle \Xi(0)(f \otimes \phi_\xi) | \Xi(0)(g \otimes \phi_\eta) \rangle\!\rangle \\
&= I_1 + I_2 + I_3 + I_4,
\end{aligned}$$
(4.15)

where

$$\begin{aligned}
I_1 &= \int_0^t \Big(\langle\!\langle (2X_1 A^\dagger(s) + X_2)(f \otimes \phi_\xi) | (g \otimes \phi_\eta) \rangle\!\rangle \\
&\quad + \langle\!\langle (f \otimes \phi_\xi) | (X_4 + 2X_5 A(s))(g \otimes \phi_\eta) \rangle\!\rangle \\
&\quad + \langle\!\langle (2X_1 A^\dagger(s) + X_2)(f \otimes \phi_\xi) | X_3(g \otimes \phi_\eta) \rangle\!\rangle\Big) d\mu_\eta(s), \\
I_2 &= \int_0^t \Big(\langle\!\langle (f \otimes \phi_\xi) | X_3(g \otimes \phi_\eta) \rangle\!\rangle + \langle\!\langle X_3(f \otimes \phi_\xi) | (g \otimes \phi_\eta) \rangle\!\rangle \\
&\quad + \langle\!\langle X_3(f \otimes \phi_\xi) | X_3(g \otimes \phi_\eta) \rangle\!\rangle\Big) d\mu_{\xi,\eta}(s), \\
I_3 &= \int_0^t \Big(\langle\!\langle (f \otimes \phi_\xi) | (2X_1 A^\dagger(s) + X_2)(g \otimes \phi_\eta) \rangle\!\rangle \\
&\quad + \langle\!\langle (X_4 + 2X_5 A(s))(f \otimes \phi_\xi) | (g \otimes \phi_\eta) \rangle\!\rangle \\
&\quad + \langle\!\langle X_3(f \otimes \phi_\xi) | (2X_1 A^\dagger(s) + X_2)(g \otimes \phi_\eta) \rangle\!\rangle\Big) d\overline{\mu_\xi}(s), \\
I_4 &= \int_0^t \Big(\langle\!\langle (2X_1 A^\dagger(s) + X_2)(f \otimes \phi_\xi) | (2X_1 A^\dagger(s) + X_2)(g \otimes \phi_\eta) \rangle\!\rangle \\
&\quad + \langle\!\langle X_6(f \otimes \phi_\xi) | (g \otimes \phi_\eta) \rangle\!\rangle \\
&\quad + \langle\!\langle (f \otimes \phi_\xi) | X_6(g \otimes \phi_\eta) \rangle\!\rangle\Big) ds.
\end{aligned}$$

From (4.15), as a set of sufficient conditions under which (4.14) holds, we have the following: for any $s \geq 0$, $\xi, \eta \in \mathcal{H}$,

$$0 = \langle\!\langle (2X_1 A^\dagger(s) + X_2)(f \otimes \phi_\xi) \,|\, (g \otimes \phi_\eta) \rangle\!\rangle$$
$$+ \langle\!\langle (f \otimes \phi_\xi) \,|\, (X_4 + 2X_5 A(s))(g \otimes \phi_\eta) \rangle\!\rangle$$
$$+ \langle\!\langle (2X_1 A^\dagger(s) + X_2)(f \otimes \phi_\xi) \,|\, X_3(g \otimes \phi_\eta) \rangle\!\rangle,$$

$$0 = \langle\!\langle (f \otimes \phi_\xi) \,|\, X_3(g \otimes \phi_\eta) \rangle\!\rangle + \langle\!\langle X_3(f \otimes \phi_\xi) \,|\, (g \otimes \phi_\eta) \rangle\!\rangle$$
$$+ \langle\!\langle X_3(f \otimes \phi_\xi) \,|\, X_3(g \otimes \phi_\eta) \rangle\!\rangle,$$

$$0 = \langle\!\langle (f \otimes \phi_\xi) \,|\, (2X_1 A^\dagger(s) + X_2)(g \otimes \phi_\eta) \rangle\!\rangle$$
$$+ \langle\!\langle (X_4 + 2X_5 A(s))(f \otimes \phi_\xi) \,|\, (g \otimes \phi_\eta) \rangle\!\rangle$$
$$+ \langle\!\langle X_3(f \otimes \phi_\xi) \,|\, (2X_1 A^\dagger(s) + X_2)(g \otimes \phi_\eta) \rangle\!\rangle,$$

$$0 = \langle\!\langle (2X_1 A^\dagger(s) + X_2)(f \otimes \phi_\xi) \,|\, (2X_1 A^\dagger(s) + X_2)(g \otimes \phi_\eta) \rangle\!\rangle$$
$$+ \langle\!\langle X_6(f \otimes \phi_\xi) \,|\, (g \otimes \phi_\eta) \rangle\!\rangle$$
$$+ \langle\!\langle (f \otimes \phi_\xi) \,|\, X_6(g \otimes \phi_\eta) \rangle\!\rangle$$

and so

$$0 = 2X_1 A^\dagger(s) + X_2 + X_4^\dagger + 2A^\dagger(s)X_5^\dagger + X_3^\dagger(2X_1 A^\dagger(s) + X_2),$$
$$0 = X_3 + X_3^\dagger + X_3^\dagger X_3,$$
$$0 = 2A(s)X_1^\dagger + X_2^\dagger + X_4 + 2X_5 A(s) + (2A(s)X_1^\dagger + X_2^\dagger)X_3,$$
$$0 = (2A(s)X_1^\dagger + X_2^\dagger)(2X_1 A^\dagger(s) + X_2) + X_6 + X_6^\dagger,$$

in which the first and third equations are equivalent. Therefore, as a set of sufficient conditions under which (4.14) holds, we have the following equations of operators:

$$0 = 2X_1 A^\dagger(s) + X_2 + X_4^\dagger + 2A^\dagger(s)X_5^\dagger + X_3^\dagger(2X_1 A^\dagger(s) + X_2),$$
$$0 = X_3 + X_3^\dagger + X_3^\dagger X_3,$$
$$0 = (2A(s)X_1^\dagger + X_2^\dagger)(2X_1 A^\dagger(s) + X_2) + X_6 + X_6^\dagger,$$

from which we have and

$$0 = X_2 + X_4^\dagger + X_3^\dagger X_2, \tag{4.16}$$
$$0 = X_1 + X_5^\dagger + X_3^\dagger X_1, \tag{4.17}$$
$$0 = X_3 + X_3^\dagger + X_3^\dagger X_3, \tag{4.18}$$
$$0 = X_6 + X_6^\dagger + (2A(s)X_1^\dagger + X_2^\dagger)(2X_1 A^\dagger(s) + X_2). \tag{4.19}$$

For a sufficient family of conditions under which (4.16)–(4.19) hold, we assume that

Assumption (U):

(**U1**) $X_3 = S - 1$ for an isometry S in $\mathcal{B}(\mathfrak{h})$,
(**U2**) $X_1 = F$ and $X_5 = -F^\dagger S$ for a bounded operator F on $\mathfrak{h}$,
(**U3**) $X_2 = G$ and $X_4 = -G^\dagger S$ for a bounded operator G on $\mathfrak{h}$,
(**U4**) $X_6 = iH - \frac{1}{2}(2A(s)F^\dagger + G^\dagger)(2FA^\dagger(s) + G)$ for a bounded selfadjoint operator H on $\mathfrak{h}$.

Therefore, we study the following quantum stochastic differential equation:

$$dU = U\left(Fd\Delta^\dagger + GdA^\dagger + (S-1)d\Lambda - G^\dagger S dA - F^\dagger S d\Delta + X_6 dt\right),$$
$$U(0) = 1 \tag{4.20}$$

with X_6 given as in (**U4**).

Proposition 4.2: *There exists a unique isometric operator-valued regular* $(\mathfrak{h}, \mathcal{H})$*-adapted process* $\{\Xi_t\}_{t\geq 0}$ *such that* (4.20) *holds*.

Proof: The proof is a simple modification of Proposition 26.2 in [23]. By Theorem 4.1 there exists a unique regular $(\mathfrak{h}, \mathcal{H})$-adapted process Ξ satisfying (4.20) and (4.6). For notational convenience, we put

$$Z_t = \Xi(t)^\dagger \Xi(t) - I, \qquad t \geq 0.$$

Then by applying Theorem 3.4, (3.17) and (4.15), we obtain that

$$\langle\!\langle f \otimes \phi_\xi | Z_t(g \otimes \phi_\eta)\rangle\!\rangle = \int_0^t \langle\!\langle f \otimes \phi_\xi | W_1(g \otimes \phi_\eta)\rangle\!\rangle \, d\mu_\eta(s)$$
$$+ \int_0^t \langle\!\langle f \otimes \phi_\xi | W_2(g \otimes \phi_\eta)\rangle\!\rangle \, \overline{d\mu_\xi}(s)$$
$$+ \int_0^t \langle\!\langle f \otimes \phi_\xi) | W_3(g \otimes \phi_\eta)\rangle\!\rangle \, d\mu_{\xi,\eta}(s)$$
$$+ \int_0^t \langle\!\langle f \otimes \phi_\xi) | W_4(g \otimes \phi_\eta)\rangle\!\rangle \, ds,$$

where

$$W_1 = Y_1^\dagger Z_s(1 + Y_2) + Z_s Y_3, \qquad W_2 = (1 + Y_2^\dagger) Z_s Y_1 + Y_3^\dagger Z_s,$$
$$W_3 = Y_2^\dagger Z_s + Z_s Y_2 + Y_2^\dagger Z_s Y_2, \qquad W_4 = Y_4^\dagger Z_s + Z_s Y_4 + Y_1^\dagger Z_s Y_1,$$

and Y_i are given as in (4.4) with constants X_i given as in (**U1**)–(**U4**) in **Assumption (U)**, which implies that

$$Z_t = \int_0^t W_1 d\mu_\eta(s) + \int_0^t W_2 \overline{d\mu_\xi}(s) + \int_0^t W_3 d\mu_{\xi,\eta}(s) + \int_0^t W_4 ds. \tag{4.21}$$

Then for any given $n \geq 1$, by iterating the integral equation (4.21) for n-times and using (4.6), we prove that for any $f, g \in \mathfrak{h}$ and $\xi, \eta \in \mathcal{H}$, there exist constants $C, K \geq 0$ such that

$$\sup_{0 \leq t \leq T} \sup_{|f| \leq 1, |g| \leq 1} |\langle f \otimes \phi_\xi | Z_t(g \otimes \phi_\eta) \rangle| \leq K \frac{C^n}{n!},$$

which implies that

$$\langle\!\langle \Xi(t)(f \otimes \phi_\xi) | \Xi(t)(g \otimes \phi_\eta) \rangle\!\rangle = \langle\!\langle f \otimes \phi_\xi | g \otimes \phi_\eta \rangle\!\rangle, \qquad t \geq 0,$$

and so $\Xi(t)$ can be extended to an isometry of $\mathfrak{H}$. $\qquad\square$

Theorem 4.3: *Let F, G, S, H be bounded operators on $\mathfrak{h}$ such that S is unitary and H is a selfadjoint operator. Then there exists a unique unitary operator-valued $(\mathfrak{h}, \mathcal{H})$-adapted regular process $U = \{U(t) \,|\, t \geq 0\}$ satisfying that*

$$dU = U \big[F d\Delta^\dagger + G dA^\dagger + (S-1) d\Lambda - G^\dagger S dA - F^\dagger S d\Delta$$
$$+ \Big(iH - \frac{1}{2}(2A(s)F^\dagger + G^\dagger)(2FA^\dagger(s) + G) \Big) dt \big], \qquad U(0) = 1.$$
$$(4.22)$$

Proof: By Proposition 4.2 there exists an isometric operator-valued adapted process U satisfying the equation (4.22). Then the adjoint operators $U(t)^\dagger$ of the operators $U(t)$ satisfies the equation:

$$dU^\dagger = \big[-S^\dagger F d\Delta^\dagger - S^\dagger G dA^\dagger + (S^\dagger - 1) d\Lambda + G^\dagger dA + F^\dagger d\Delta$$
$$+ \Big(-iH - \frac{1}{2}(2A(s)F^\dagger + G^\dagger)(2FA^\dagger(s) + G) \Big) dt \big] U^\dagger, \quad U^\dagger(0) = 1.$$
$$(4.23)$$

Then by the quantum Itô formula described as in (3.17), we obtain that

$$d\,(UU^\dagger) = U \big[-S^\dagger F d\Delta^\dagger - S^\dagger G dA^\dagger + (S^\dagger - 1) d\Lambda + G^\dagger dA + F^\dagger d\Delta$$
$$+ \Big(-iH - \frac{1}{2}(2AF^\dagger + G^\dagger)(2FA^\dagger + G) \Big) dt \big] U^\dagger$$
$$+ U \big[F d\Delta^\dagger + G dA^\dagger + (S-1) d\Lambda - G^\dagger S dA - F^\dagger S d\Delta$$
$$+ \Big(iH - \frac{1}{2}(2AF^\dagger + G^\dagger)(2FA^\dagger + G) \Big) dt \big] U^\dagger$$
$$+ U \big[\{ -(S-1)S^\dagger(2A^\dagger F + G) \} dA^\dagger + \{ (S-1)(S^\dagger - 1) \} d\Lambda$$
$$+ \{ -(G^\dagger + 2F^\dagger A)S(S^\dagger - 1) \} dA + (G^\dagger + 2F^\dagger A)(G + 2FA^\dagger) dt \big] U^\dagger$$
$$= 0.$$

Since $U(0) = 1$, it follows that $U(t)U(t)^\dagger = 1$ for any $t \geq 0$, and so $U(t)$ is coisometry for any $t \geq 0$. Therefore, $U(t)$ is unitary for any $t \geq 0$. □

Corollary 4.4: *Let F, G, S, H be bounded operators on $\mathfrak{h}$ such that S is unitary and H is a selfadjoint operator. Then there exists a unique unitary operator-valued $(\mathfrak{h}, \mathcal{H})$-adapted regular process $U = \{U(t) \,|\, t \geq 0\}$ satisfying that*

$$dU = \left[Fd\Delta^\dagger + GdA^\dagger + (S-1)d\Lambda - G^\dagger S dA - F^\dagger S d\Delta \right.$$
$$\left. - \left(iH + \frac{1}{2}(2A(s)F^\dagger + G^\dagger)(2FA^\dagger(s) + G) \right) dt \right] U, \quad U(0) = 1.$$
$$(4.24)$$

Proof: Since the equation (4.24) is the adjoint of (4.22) by changing F, G and S in (4.22) by $-F$, $-S^\dagger G$ and $S^\dagger$, respectively, the proof is immediate from Theorem 4.3. □

Corollary 4.5: *([9]) Let G, S, H be bounded operators on $\mathfrak{h}$ such that S is unitary and H is a selfadjoint operator. Then there exists a unique unitary operator-valued $(\mathfrak{h}, \mathcal{H})$-adapted regular process $U = \{U(t) \,|\, t \geq 0\}$ satisfying that*

$$dU = U \left(GdA^\dagger + (S-1)d\Lambda - G^\dagger S dA + \left(iH - \frac{1}{2}G^\dagger G \right) dt \right), \quad U(0) = 1.$$
$$(4.25)$$

Proof: The equation (4.25) coincides with the equation (4.22) with $F = 0$. Therefore, the proof is immediate from Theorem 4.3. □

Corollary 4.6: *Let F, S, H be bounded operators on $\mathfrak{h}$ such that S is unitary and H is a selfadjoint operator. Then there exists a unique unitary operator-valued $(\mathfrak{h}, \mathcal{H})$-adapted regular process $U = \{U(t) \,|\, t \geq 0\}$ satisfying that $U(0) = 1$ and*

$$dU = U \left(Fd\Delta^\dagger + (S-1)d\Lambda - F^\dagger S d\Delta + \left(iH - 2F^\dagger A_s A_s^\dagger F \right) dt \right). \quad (4.26)$$

Proof: The equation (4.26) coincides with the equation (4.22) with $G = 0$. Therefore, the proof is immediate from Theorem 4.3. □

Remark 4.7: Consider the following quantum stochastic differential equation:

$$dU_{s,t} = U_{s,t} \big[Fd\Delta^\dagger + GdA^\dagger + (S-1)d\Lambda - G^\dagger SdA - F^\dagger Sd\Delta$$
$$+ \Big(iH - \frac{1}{2}(2A(t)F^\dagger + G^\dagger)(2FA^\dagger(t) + G) \Big) dt \big],$$

$$U_{s,s}(0) = 1, \qquad 0 \le s \le t. \tag{4.27}$$

Then by Theorem 4.3, there exists a unique unitary operator-valued $(\mathfrak{h}, \mathcal{H})$-adapted regular process $U = \{U_{s,t} \,|\, t \ge s \ge 0\}$ satisfying (4.27). The vacuum expectation $E[U_{s,t}]$ of the unitary solution $U_{s,t}$ satisfies the differential equation:

$$\frac{dV_{s,t}}{dt} = V_{s,t} \left(\Big(-\frac{1}{2}G^*G + iH \Big) - 2F^\dagger Ft \right) - R_{s,t}G^\dagger F, \qquad U_{s,s} = 1,$$

where $R_{s,t} = E[U_{s,t}A_t^\dagger]$. If $G = 0$, then the vacuum expectation $E[U_{s,t}]$ of the unitary solution $U_{s,t}$ satisfies the differential equation:

$$\frac{dV_{s,t}}{dt} = V_{s,t} \left(iH - 2F^\dagger Ft \right), \qquad U_{s,s} = 1. \tag{4.28}$$

Therefore, the unique unitary solution $\{U_{s,t} | t \ge s \ge 0\}$ of the quantum stochastic differential equation (4.27) with $G = 0$ is a unitary dilation of the evolution $\{V_{s,t} | t \ge s \ge 0\}$ given by the differential equation (4.28).

References

1. L. Accardi, A. Boukas and H.-H. Kuo: *On the unitarity of stochastic evolutions driven by the square of white noise*, Infin. Dimens. Anal. Quantum Probab. Relat. Top. **4** (2001), 579–588.
2. V. P. Belavkin: *Quantum stochastic calculus and quantum nonlinear filtering*, J. Multivariate Anal. **42** (1992), 171–201.
3. V. P. Belavkin: *Quantum stochastic positive evolutions: characterization, construction, dilation*, Comm. Math. Phys. **184** (1997), 533–566.
4. B. V. R. Bhat and K. B. Sinha: *A stochastic differential equation with time-dependent and unbounded operator coefficients*, J. Funct. Anal. **114** (1993), 12–31.
5. D. M. Chung, U. C. Ji and N. Obata: *Higher powers of quantum white noises in terms of integral kernel operators*, Infin. Dimen. Anal. Quantum Probab. Rel. Top. **1** (1998), 533–559.
6. D. M. Chung, U. C. Ji and N. Obata: *Quantum stochastic analysis via white noise operators in weighted Fock space*, Rev. Math. Phys. **14** (2002), 241–272.
7. D. Goswami and K. B. Sinha: *Hilbert modules and stochastic dilation of a quantum dynamical semigroup on a von Neumann algebra*, Comm. Math. Phys. **205** (1999), 377–403.

8. Z.-Y. Huang: *Quantum white noises-white noise approach to quantum stochastic calculus,* Nagoya Math. J., **129** (1993), 23–42.

9. R. L. Hudson and K. R. Parthasarathy: *Quantum Itô's formula and stochastic evolutions,* Commun. Math. Phys. **93** (1984), 301–323.

10. R. L. Hudson and R. F. Streater: *Itô's formula is the chain rule with Wick ordering,* Phys. Lett. **86A** (1981), 277–284.

11. U. C. Ji: *Stochastic integral representation theorem for quantum semimartingales,* J. Funct. Anal. **201** (2003), 1–29.

12. U. C. Ji and N. Obata: *Quantum white noise calculus,* in "Non-Commutativity, Infinite-Dimensionality and Probability at the Crossroads (N. Obata, T. Matsui and A. Hora, Eds.)," pp. 143–191, World Scientific, 2002.

13. U. C. Ji and N. Obata: *Annihilation-derivative, creation-derivative and representation of quantum martingales,* Comm. Math. Phys. **286** (2009), 751–775.

14. U. C. Ji and N. Obata: *Quantum stochastic gradients,* Interdiscip. Inform. Sci. **14** (2009), 345–359.

15. U. C. Ji and K. B. Sinha: *Quantum stochastic calculus associated with quadratic quantum noises,* J. Math. Phys. **57** (2016), 022702.

16. G. Lindblad: *On the generators of quantum dynamical semigroups,* Comm. Math. Phys. **48** (1976), 119–130.

17. J. M. Lindsay: *Quantum and non-causal stochastic calculus,* Probab. Th. Rel. Fields **97** (1993), 65–80.

18. P. A. Meyer: "Quantum Probability for Probabilists," Lect. Notes in Math. **1538**, Springer–Verlag, 1993.

19. A. Mohari and K. B. Sinha: *Quantum stochastic flows with infinite degrees of freedom and countable state Markov processes,* Sankhyā Ser. A **52** (1990), 43–57.

20. N. Obata: *Generalized quantum stochastic processes on Fock space,* Publ. RIMS **31** (1995), 667–702.

21. N. Obata: *Quantum stochastic differential equations in terms of quantum white noise,* Nonlinear Analysis, Theory, Methods and Applications **30** (1997), 279–290.

22. N. Obata: *Wick product of white noise operators and quantum stochastic differential equations,* J. Math. Soc. Japan. **51** (1999), 613–641.

23. K. R. Parthasarathy: "An Introduction to Quantum Stochastic Calculus," Birkhäuser, 1992.

24. K. B. Sinha and D. Goswami: "Quantum Stochastic Processes and Non-commutative Geometry," Cambridge Tracts in Mathematics **169**, Cambridge University Press, Cambridge, 2007.

ITÔ FORMULA FOR GENERALIZED REAL AND COMPLEX WHITE NOISE FUNCTIONALS

Yuh-Jia Lee

Department of Applied Mathematics
National University of Kaohsiung
Kaohsiung, Taiwan 811
yuhjialee@gmail.com

It follows immediately from the definition of the composition $f(B(t))$ of f with the Brownian motion $B(t)$, we are able to derive "Itô formula" given in the following form

$$f(B(b)) - f(B(a)) = \int_a^b \partial_t^* f'(B(s)) \, ds + \frac{1}{2} \int_a^b f''(B(s)) \, ds$$

without using the Itô integral, where f may be a smooth function or a tempered distribution. Compare with the "classical" Itô formula, the term $\int_a^b \partial_t^* f'(B(s)) \, ds$ should be realized as the Itô integral $\int_a^b f(B(s)) \, dB(s)$. The integral $\int_a^b \partial_t^* X(t) \, dt$, also known as the Hitsuda-Skorokhod integral, becomes a natural extension of the Itô type stochastic integral which is defined for non-adapted process $X(t)$ as well. In this chapter the Itô formula for complex Brownian and its connection with the Itô formula (real) Brownian motion are also discussed.

1. Introduction

In this chapter we first derive the "Itô's formula" without the definition of Itô integral via white noise analysis, in the proof we show that the Itô integral

$$\int_a^b f(B(t)) \, dB(t)$$

coincide with the the Hitsuda-Skorokhod integral

$$\int_a^b \partial_t^* f(B(t)) \, dt.$$

The above result has been extended by Kubo and Takanaka [4] from $f(B(t))$ to general adapted integrand $X(t)$. It is natural to ask what is the definition of Hitsuda-Skorokhod integral in terms of the Itô type Riemann sum when the integrand is nonadpated. Partial results has been done in the case of stochastic integral with respect to Brownian bridge. The stochastic integral and Itô formula for complex Brownian motion will also be discussed in this chapter. It is shown that the calculus of Itô integral for complex Brownian motion follows the rule of that of Stratonovich integral. Using the advantage of the Stratonovich integral and the associated Itô formula, we show that the Itô formula for real Brownain motion can be recovered by the Itô formula of complex Brownian motion (see [12]).

2. Generalized White Noise Functionals

To start with we list in the following the basic notations which will be used in this chapter.

- S : the Schwartz space on $\mathbb{R}$
- S': the space of tempered distribution
- $(\cdot, \cdot)$: the S'-S pairing
- $S_0 = L^2(\mathbb{R})$
- A : $Au = -u'' + 1 + u^2$, A is densely defined in S_0

- $\{e_j : j = 0, 1, 2. \dots\}$: CONS of S_0, consisting of eigenfunctions of A with corresponding eigenvalues $\{2j + 2 : j = 0, 1, 2, \dots\}$
  Let $S_p = \{f \in S' : \|f\|_p < \infty\}$ where

$$\|f\|_p^2 = \sum_{j=0}^{\infty}(2j + 2)^p(f, e_j)^2.$$

- $S = \cap_{p \geq 0} S_p;$ $S' = \cup_{p \geq 0} S_{-p}$
- $S \subset H \subset S'$ forms a Gel'fand triple.
- μ: a standard Gaussian measure defined on $(S', \mathcal{B}(S'))$ with the characteristic functional $\mathcal{C}$ on S given by

$$\mathcal{C}(\eta) = \int_{S'} e^{(x,\eta)}\mu(dx) = e^{-\frac{1}{2}\|\eta\|_0^2}$$

where $\|\eta\|_0 = \left\{\int_{-\infty}^{+\infty} \eta(t)^2\, dt\right\}^{1/2}$ $(\eta \in S)$.
- $(L^2) := L^2(S', \mu)$
- $\mu f := \mu * f$.

–An Equivalent Scheme of Hida Calculus

We adapt a new but equivalent scheme of the the Hida calculus introduced in Lee [7] (see also [2, 5, 6, 8, 9, 12, 13]). Let $\mathcal{A}_\infty = \cap_{p>0}\mathcal{A}_p$. Endow $\mathcal{A}_\infty$ with the projective topology. Then $\mathcal{A}_\infty$ becomes a topological space which serves as the test functionals. Let $\mathcal{A}_p^*$ and $\mathcal{A}_\infty^*$ be respectively the dual space of $\mathcal{A}_p$ and $\mathcal{A}_\infty$. Then $\mathcal{A}_\infty^*$ becomes the inductive limit of $\mathcal{A}_p^*$. Members of $\mathcal{A}_\infty^*$ are called generalized white noise functionals or simply, Hida distribution. Moreover we have the following densely inclusion relations:

$$\mathcal{A}_\infty \subset \mathcal{A}_p \subset \mathcal{A}_q \subset (L^2) \subset\subset \mathcal{A}_q^* \subset \mathcal{A}_p^* \subset \mathcal{A}_\infty^*,$$

and the triple $(\mathcal{A}_\infty \subset (L^2) \subset \mathcal{A}_\infty^*)$ form a Gel'fand triple. The basic properties of the test functionals are given below.

Proposition 2.1: *Let $f \in \mathcal{A}_\infty$. Then we have*

(a) *For $h_1, \ldots, h_n \in \mathcal{S}$ and for $p \in \mathbb{N}$,*

$$|D^n f(z)h_1 \cdots h_n| \leq \|f\|_{\mathcal{A}_p} \exp\left[\|z\|_{-p}^2 \left(\sum_{j=1}^\infty \|h_j\|_{-p}\right)^2\right].$$

(b) $\sum_{n=0}^\infty \frac{1}{n!}D^n f(0)z^n$ *converges to f in $\mathcal{A}_\infty$ for any $f \in \mathcal{A}_\infty$.*
(c) $\mathcal{A}_\infty$ *is an algebra.*
(d) *The Wiener–Itô decomposition of $f \in \mathcal{A}_\infty$ converges to f in $\mathcal{A}_\infty$.*
(e) *For $f \in \mathcal{A}_\infty$, define $\mathcal{F}_{\alpha,\beta}f(y) = \int_{\mathcal{S}^*} f(\alpha x + \beta y)\mu(dx)$ for $\alpha, \beta \in \mathbb{C}$. Then $\mathcal{F}_{\alpha,\beta}(\mathcal{A}_\infty) \subset \mathcal{A}_\infty$ and $\mathcal{F}_{\alpha,\beta}$ is continuous on $\mathcal{A}_\infty$.*

Given $F \in \mathcal{A}_\infty^*$, we define the S-transform of F as follows:

Definition 2.2:

$$SF(\xi) = \begin{cases} \mu * F(\xi), & \text{if } F \in L^2[\mathcal{S}', \mu]; \\ e^{-\frac{1}{2}|\xi|^2}\langle\!\langle F, e^{(\cdot, \xi)}\rangle\!\rangle, & \text{if } F \in (\mathcal{S})^*, \end{cases}$$

where $\xi \in \mathcal{S}$.

SF is also denoted by U_F, U_F is called the U-functional F.

The following theorem gives a criterion for convergence of generalized white noise functionals. We state it without proof, for details we refer the reader to [7].

Theorem 2.3: *Let $\{\Phi_n, \Phi_0\}$ be a sequence in $\mathcal{A}_\infty^*$ and $f_n = U_{\Phi_n}$, $f_0 = U_{\Phi_0}$. Then the following are equivalent:*

(i) $\Phi_n \to \Phi_0$ *in* $\mathcal{A}_\infty^*$.
(iii) $\exists q \in \mathbb{Z}$ *such that* $f_n \to f_0$ *in* $\mathcal{A}_q$.
(iv) $\exists r \in \mathbb{Z}, \exists C_r$ *such that for all* $\xi \in \mathcal{S}_r$.

$$|f_n(\xi)| \le C_r \exp\left[\frac{1}{2}\|\xi\|_r^2\right] \tag{2.1}$$

for all $n \in \mathbb{N}_0$, *and* $f_n(\xi) \to f_0(\xi)$.

3. Itô Formula for White Noise Functionals

Let (H, B) be an abstract Wiener space with abstract Wiener measure $\mu = p_1$. Let B^* be the dual space which is regarded as the subspace of H.
Let $\xi \in B^*$. Define

$$\widetilde{\xi}(x) = (x, \xi).$$

Then $\widetilde{\xi} \in \mathcal{A}_\infty$ and $\widetilde{\xi}$ is normal distributed with mean zero and variance $\|\xi\|_H^2$.

For any $h \in H$, there exists a sequence $(\xi_n) \subset B^*$ such that $|\xi_n - h|_H \to 0$. It follows that $\int_B |\widetilde{\xi}_n - \widetilde{\xi}_m|^2 \mu(dx) = \|\xi_n - \xi_m\|_H^2 \to 0$ as $n, m \to \infty$. Thus $\{\widetilde{\xi}_n\}$ forms a Cauchy sequence in $L^2(B)$ so that the $L^2(B)$-limit of $\{\widetilde{\xi}^n\}$ exists. Define

$$\widetilde{h} = L^2(B) - \lim_{n\to\infty} \widetilde{\xi}.$$

Then $\widetilde{h} \sim N(0, \|h\|_H^2)$. In notation, we also write

$$\widetilde{h}(x) = \langle x, h \rangle.$$

When $H = L^2(\mathbb{R})$, we consider $(L^2(\mathbb{R}), \mathcal{S}')$ as the union of the abstract Wiener spaces $(L^2(\mathbb{R}), \mathcal{S}_p)$. Then $\widetilde{h}$ is well-defined as a normal distributed random variable with mean 0 and variance $|h|_0^2$.

The Brownian motion $B(t)$ on the probability space $(\mathcal{S}', \mathcal{B}(\mathcal{S}), \mu)$ may be represented by

$$B(t, x) = \begin{cases} \langle x, 1_{(0,t]} \rangle, & t \ge 0 \\ -\langle x, 1_{(t,0]} \rangle, & t < 0, \end{cases}$$

for almost all $x \in \mathcal{S}'$.
Let

$$h_t = \begin{cases} 1_{(0,t]}, & t \ge 0 \\ -1_{(t,0]}, & t < 0, \end{cases}$$

then

$$B(t, x) = \langle x, h_t \rangle.$$

For any test functional φ, we have

$$\langle\langle \dot{B}(t), \varphi \rangle\rangle = \frac{d}{dt} \langle\langle B(t), \varphi \rangle\rangle$$

$$= \lim_{\epsilon \to 0} \int_{S'} \frac{1}{\epsilon} \langle x, h_{t+\epsilon} - h_t \rangle \varphi(x) \mu(dx)$$

$$= \lim_{\epsilon \to 0} D\mu\varphi(0) \left\{ \frac{1}{\epsilon}(h_{t+\epsilon} - h_t) \right\}$$

$$= D\mu\varphi(0)\delta_t.$$

It is easy to see that the mapping $\varphi \to D\mu\varphi(0)\delta_t$ is continuous on $\mathcal{A}_\infty$. This leads to the definition of white noise given as follows

$$\langle\langle \dot{B}(t), \varphi \rangle\rangle = D\mu\varphi(0)\delta_t.$$

To drive the Itô formula we start with defining the composition of generalized function with Brownian motion. Given $f \in \mathcal{S}(\mathbb{R}^n)$ and $h_i \in L^2$, $i = 1, 2, 3 \ldots$, then by Fourier inversion formula, we have

$$f(\widetilde{h}_1, \widetilde{h}_2, \ldots, \widetilde{h}_n) = \frac{1}{2\pi^n} \int_{\mathbb{R}^n} \widehat{f}(u_1, \ldots, u_n) e^{i \sum_{j=1}^n u_i \widetilde{h}_i} \, du_1 \ldots du_n.$$

It follows by elementary calculus on abstract Wiener space that we obtain

$$\langle\langle f(\widetilde{h}_1, \widetilde{h}_2, \ldots, \widetilde{h}_n), \varphi \rangle\rangle = (f, G_{\mathbf{h}, \varphi}), \tag{3.1}$$

where

$$G_{\mathbf{h}, \varphi}(\mathbf{u}) = (1/\sqrt{2\pi})^n \mathcal{F}_{1,i} \varphi([\mathbf{u}, \mathbf{h}]) \exp(-\frac{1}{2} \| [\mathbf{u}, \mathbf{h}] \|_0^2),$$

for $\varphi \in \mathcal{A}_\infty$, where $\mathbf{u} = (u_1, u_2, \ldots, u_n)$, $\mathbf{h} = (h_1, h_2, \ldots, h_n)$ and $[\mathbf{u}, \mathbf{h}] = \sum_{j=1}^n u_i h_i$. Observe that the mapping

$$\varphi \to G_{\mathbf{h}, \varphi}$$

is continuous from $\mathcal{A}_\infty$ to $\mathcal{S}$ so that the formula (3.1) make sense for $f \in \mathcal{S}'$ as well. Further more, it is easy to see that the mapping $\varphi \to (f, G_{\mathbf{h}, \varphi})$ define a continuous linear functionals on $\mathcal{A}_\infty$. This leads to the following

Definition 3.1: Let $f \in \mathcal{S}'(\mathbb{R}^n)$ and $h_i \in L^2$, $i = 1, 2, 3 \ldots$. For $\varphi \in \mathcal{A}_\infty$ define

$$\langle\langle f(\widetilde{h}_1, \widetilde{h}_2, \ldots, \widetilde{h}_n), \varphi \rangle\rangle = (f, G_{\mathbf{h}, \varphi}).$$

Then $f(\widetilde{h}_1, \widetilde{h}_2, \ldots, \widetilde{h}_n)$ define a generalized Brownian functional.

For example, the Donsker's delta function $\delta_x(B(t))(t>0)$ may be defined by

$$\langle\!\langle \delta_x(\widetilde{h}_t), \varphi \rangle\!\rangle := \frac{1}{2\pi} \int_{-\infty}^{\infty} e^{-ias-\frac{1}{2}s^2 t} \mathcal{F}_{1,i}\varphi(sh_t)ds$$

$$:= \frac{1}{2\pi} \int_{-\infty}^{\infty} e^{-ias-\frac{1}{2}s^2 t} \left\{ \int_{\mathcal{S}'} \varphi(y + i\, sh_t)\mu(dy) \right\} ds.$$

Now we are ready to derive the Itô formula.

For $f \in \mathcal{S}'$, it follows from Definition 3.1 that the generalized functional $f(B(t)) = f(\widetilde{h}_t)$ for $t > 0$ is given by

$$\langle\!\langle f(B(t)), \varphi \rangle\!\rangle := (f, \widehat{G}_{t,\varphi})$$

where $\widehat{G}_{t,\varphi}(u) = (1/\sqrt{2\pi})\mathcal{F}_{1,i}\varphi(u1_{(0,t]}) \exp(-\frac{1}{2}u^2 t)$.

If we differentiate $f(B(t))$ with respect to t we immediately obtain:

$$\frac{d}{dt}\langle\!\langle f(B(t)), \varphi \rangle\!\rangle$$

$$= (f_{[u]}, iu\left\{ 1/\sqrt{2\pi}\mathcal{F}_{1,i}\partial_t\varphi(u1_{(0,t]})e^{-\frac{1}{2}u^2 t} \right\}$$

$$+ (f_{[u]}, -\frac{1}{2}u^2\left\{ (1/\sqrt{2\pi})\mathcal{F}_{1,i}\partial_t\varphi(u1_{(0,t]})e^{-\frac{1}{2}u^2 t} \right\}$$

$$= \langle\!\langle f(B(t)), \partial_t\varphi \rangle\!\rangle + \langle\!\langle \frac{1}{2}f''(B(t)), \varphi \rangle\!\rangle.$$

The generalized Itô's formula follows:

$$\frac{d}{dt}f(B(t)) = \partial_t^* f'(B(t)) + \frac{1}{2}f''(B(t)).$$

It can be shown that

$$\int_a^b \partial_t^* f'(B(t))dt = \int_a^b f'(B(t))dB(t).$$

If one replace the Brownian motion by any other normal processes

$$X_t(x) = \langle x, \beta_t \rangle,$$

one may derive a new Itô formula by differentiating $f(X_t)$ with respect to t. To demonstrate this method, we give an example as follow.

Example 3.2: Define

$$\langle\!\langle f(B(t), B(1), \varphi \rangle\!\rangle) = (f, \widehat{H}_{t,\varphi}),$$

where

$$\widehat{H}_{t,\varphi} = \frac{1}{\sqrt{2\pi}^2} \mathcal{F}_{1,i}\varphi(u\,h_t + v\,h_1)e^{-\frac{1}{2}u^2\|u\,h_t + v\,h_1\|_0^2}.$$

Differentiating with respect to t and then integrating from $a > 0$ to $b > a(1 > b)$ we obtain the Hitsuda formula (cf. Kuo [5]) as follows:

$$f(B(b), B(1)) - f(B(a), B(1)) = \int_a^b \partial_t^* f_x(B(t), B(1))\, dt$$

$$+ \int_a^b f_{xy}(B(t), B(1))dt + \frac{1}{2}\int_a^b f_{xx}(B(t), B(1))\, dt.$$

Next we give an application of Hitsuda formula.

Apply Hitsuda formula with $f(xy) = xy$, we immediately have

$$B(b)B(1) - B(a)B(1) = \int_a^b \partial_t^* B(1)\, dt + (b - a),$$

or,

$$\int_a^b \partial_t^* B(1)\, dt = (B(b) - B(a))B(1) - (b - a).$$

–Itô Formula for Non-Adapted Processes

For the more general case $f(X_t)$ with $X_t(x) = \langle x, h_t \rangle$ with $\{X_t\}$ being a normal processes (which is non-adapted generally), one may also apply the same argument above to derive the following "Itô" formula:

$$f(X(b)) = f(X(a)) + \int_a^b D_{\dot{h}_t}^* f'(B(t))\, dt + \int_a^b \{\frac{d}{dt}\|h_t\|_0\}f''(X(t))\, dt,$$

where $\dot{h}_t = \frac{d}{dt}h_t$.

Again a new integral such as $\int_a^b D_{\dot{h}_t}^* f(t)\, dt$ arises.

As an example we derive the Itô formula for Brownian Bridge.

Example 3.3: The Brownian Bridge $X(t)$ may represented by

$$X(t) = B(t) - tB(1) = \widetilde{\beta}_t = \widetilde{h}_t - t\widetilde{h}_1, \quad (\beta_t = h_t - th_1).$$

Clearly $\|\beta_t\|_0^2 = t - t^2$. Let $k_t = \frac{d}{dt}\beta_t$. Then, for $f \in \mathcal{S}'$, we have

$$f(X(b)) - f(X(a)) = \int_a^b D_{k_t}^* f'(X(t))\, dt + \int_a^b \frac{1}{2}(1 - 2t)f''(X(t))\, dt$$

which exist in the generalized sense, where $0 < a < b < 1$.

Let $\{Y_t : a \le t \le b\}$, $0 < a < b < 1$ be a continuous $(\mathcal{S})^*$-valued process, we define

$$\int_a^b Y_t \, dX(t+) := \lim_{|\Gamma| \to 0} \sum_{j=1}^{n} (\widetilde{\beta}_{t_j} - \widetilde{\beta}_{t_{j-1}}) Y_{t_{j-1}}$$

provided that the limit exist in $(\mathcal{S})^*$, where $\Gamma = \{a = t_0 < t_1 < \cdots < t_n = b\}$. Then one can show that

$$\int_a^b D_{k_t}^* f'(X(t)) \, dt = \int_a^b f'(X(t)) dX(t+) + \int_a^b t \, f''(X(t)) \, dt.$$

The above identity also give the probabilistic meaning of the stochastic integral

$$\int_a^b D_{k_t}^* f'(X(t)) \, dt.$$

Next we introduce a new integral initiated by Kuo [1].

Example 3.4: In [1], the authors define the following stochastic integral: Let $f(t)$, $a \le t \le b$, be adapted and $\varphi(t), a \le t \le b$ be instantly independent (i.e. $\varphi(t)$ is independent of $\sigma\{B(s), s \le t\}$). Define the stochastic integral of $f(t)\varphi(t)$ by

$$I(f\varphi) = \int_a^b f(t)\varphi(t) \, dB(t) = \lim_{\|\Delta\| \to 0} \sum_{i=1}^{n} f(t_{i-1})\varphi(t_i)(B(t_i) - B(t_{i-1}))$$

provided the limit exist in probability, where $\|\Delta\| = max\{|t_j - t_{j-1}|\}$.

Apply S-transform and Theorem 2.3, one can prove that

$$\int_a^b f(t)\varphi(t) \, dB(t) = \int_a^b \partial^*[f(t)\varphi(t)] \, dB(t).$$

4. Complex Brownian Functionals

By a complex Brownian functionals we mean a function of complex Brownian motion given by

$$Z(t, \omega) = B_1(t, \omega) + iB_2(t, \omega)$$

where B_1 and B_2 are independent real-valued standard Brownian motions. $Z(t)$ is normally distributed with mean zero and variance parameter $|t|$.

Let $(\mathcal{S}_c', \mathcal{B}(\mathcal{S}_c'), \nu(dz))$ be the underlying probability space, where $\mathcal{S}$ is the Schwartz space with dual space $\mathcal{S}'$, $\mathcal{S}_c'$ is the complexification of $\mathcal{S}_c'$ which is identified as the product space $\mathcal{S}' \times \mathcal{S}'$, $\mathcal{B}(\mathcal{S}_c')$ the Borel field of

$S' \times S'$ and $\nu(dz)$ denotes the product measure $\mu_{1/2}(dx)\mu_{1/2}(dy)$, where μ_t denotes the Gaussian measure defined on S' with characteristic function given by

$$C(\xi) = \int_{S'} e^{i(x,\xi)} \mu_t(dx) = e^{-t|\xi|^2/2}.$$

The complex Brownian motion on $(S'_c, \mathcal{B}(S'_c), \nu(dz))$ may be represented by

$$Z_t(x,y) = \langle x, h_t \rangle + i\langle y, h_t \rangle,$$

where

$$h_t = \begin{cases} 1_{(0,t]}, & t > 0, \\ -1_{[t,0]}, & t < 0. \end{cases}$$

The calculus of complex Brownian functional is then performed with respect to the measure $\mu(dz)$. For example, let $f : \mathbb{C} \to \mathbb{C}$ be an entire function of exponential growth. Then we have

$$E[|f(Z(t))|^2] = \int_{S'} \int_{S'} |f(\langle x + iy, h_t \rangle)|^2 \, \mu_{1/2}(dx)\mu_{1/2}(dy).$$

The above identity gives a connection between the function of complex Brownian motion and the Segal-Bargmann entire functionals.

–Itô Formula for Entire Brownian Functionals

We shall show that, for any Segal-Bargmann entire function F, the Itô formula is given by

$$F(Z(b)) - F(Z(a)) = \int_a^b F'(Z(t))dZ(t).$$

Definition 4.1: A single-valued function f defined on H_c is called a Segal-Bargmann entire function if it satisfies the following conditions:

(i) f is analytic in H_c.
(ii) The number

$$M_f := \sup_P \int_H \int_H |f(Px + iPy)|^2 n_t(dx)n_t(dy)$$

is finite, where n_t denoted as the Gaussian cylinder measure on H with variance parameter $t > 0$ and P's run through all orthogonal projections on H.

Denote the class of Segal-Bargmann entire function on H by $\mathcal{SB}_t[H]$ and define $\|f\|_{\mathcal{SB}_t[H]} = \sqrt{M_f}$. Then $(\mathcal{SB}_t[H], \|\cdot\|_{\mathcal{SB}_t[H]})$ is a Hilbert space.

It follows immediately from Lee [7] that we have

$$
\begin{aligned}
\|f\|^2_{\mathcal{SB}_t[H]} &= \sum_{k=0}^{\infty} \frac{(2t)^k}{k!} \left(\sum_{i_1,\dots,i_k=1}^{N} |D^k f(0) e_{i_1} \cdots e_{i_k}|^2 \right) \\
&= \sum_{k=0}^{\infty} \frac{(2t)^k}{k!} \|D^n f(0)\|^2_{HS^2[H]}.
\end{aligned}
\tag{4.1}
$$

When $t = 1/2$, we simply denote $\mathcal{SB}_t[H]$ by $\mathcal{SB}[H]$, where $\|S\|_{HS^n[H]}$ denotes the Hilbert-Schmidt norm of an n-linear operator $S \in L^n(H)$ defined by

$$
\|S\|_{HS^n(H)} := \left(\sum_{i_1,\dots,i_k=1}^{\infty} |S e_{i_1} \cdots e_{i_k}|^2 \right)^{1/2}
$$

which is independent of the choice of CONS $\{e_i\}$ of H.

–Definition of Infinite-Dimensional Segal-Bargman Entire Functionals

Definition 4.2: For each $p \in \mathbb{R}$, define

$$
|\phi|_p = \left(\sum_{n=0}^{\infty} \frac{\|D^n \phi(0)\|^2_{HS^n[S_{-p}]}}{n!} \right)^{1/2}
$$

and set

$$
\mathcal{SB}_p = \{\phi \in \mathcal{SB}[S_{-p}] : |\phi|_p < \infty\}.
$$

Let $\mathcal{SB}_\infty$ be the projective limit of $\mathcal{SB}_p$ for $p \geq 0$ and let $\mathcal{SB}'_\infty$ be the dual space of $\mathcal{SB}_\infty$. We note that

$$
\mathcal{SB}_\infty = \mathcal{A}_\infty.
$$

$\mathcal{SB}_\infty$ is a nuclear space and we have the following continuous inclusions:

$$
\mathcal{SB}_\infty \subset \mathcal{SB}_p \subset \mathcal{SB}[L^2] = \mathcal{SB} \subset \mathcal{SB}'_p \subset \mathcal{SB}'_\infty.
$$

The space $\mathcal{SB}_\infty$ will serve as test functionals and $\mathcal{SB}'_\infty$ is referred as the generalized complex Brownian functionals.

The space $\mathcal{SB}'_p$ may be identified as the space of entire functions defined on $\mathcal{S}_{p,c}$ such that : $|\phi|_{-p} < \infty$ and the pairing of $\mathcal{SB}'_\infty$ and $\mathcal{SB}_\infty$ is defined by

$$\langle\langle \Phi, \varphi \rangle\rangle = \sum_{n=0}^{\infty} \frac{1}{n!} \langle\langle D^n \overline{\Phi}(0), D^n \varphi(0) \rangle\rangle_{HS^n},$$

where

$$\langle\langle D^n \overline{\Phi}(0), D^n \varphi(0) \rangle\rangle_{HS^n}$$

$$:= \sum_{i_1,\ldots,i_n=1}^{n} \left[\overline{D^n \Phi(0) e_{i_1} \cdots e_{i_n}} D^n \varphi(0) e_{i_1} \cdots e_{i_n} \right].$$

Definition 4.3: (One-dimensional Segal-Bargman entire functions) If $\phi(z)$ can be represented by a formal power series $\sum_{n=0}^{\infty} a_n z^n$, we define

$$|\phi|_p = \left(\sum_{n=0}^{\infty} (2n+2)^{2p} n! |a_n|^2 \right)^{1/2}$$

and let

$$\mathcal{SB}_p(\mathbb{R}) = \{\phi : |\phi|_p < \infty\}.$$

If $\phi(z)$ is a formal power series represented by $\sum_{n=0}^{\infty} b_n z^n$, we define

$$|\phi|_{-p} = \left(\sum_{n=0}^{\infty} n! |b_n|^2 (2n+2)^{-2p} \right)^{1/2}.$$

Then the dual space $\mathcal{SB}'_p(r)$ of $\mathcal{SB}_p(r)$ is characterized by

$$\mathcal{SB}_{-p}(\mathbb{R}) = \{\phi : |\phi|_{-p} < \infty\}$$

The space $\mathcal{SB}_\infty[\mathbb{R}]$ is defined as the projective limit of $\mathcal{SB}_p[\mathbb{R}]$ with dual space $\mathcal{SB}'_\infty[\mathbb{R}] = \bigcup_{p>0} \mathcal{SB}'_p[\mathbb{R}]$.

Definition 4.4: (Composition of generalized function with complex Brownian motion) Let $\psi \in \mathcal{SB}_\infty$. Then, for any one dimensional generalized Segal-Bargman entire function $f \in \mathcal{SB}'_\infty(\mathbb{R})$, represented by $\psi(z) = \sum_{n=0}^{\infty} a_n z^n$, we have

$$\langle\langle f(Z(t)), \psi \rangle\rangle_c = \sum_{n=0}^{\infty} b_n D^n \psi(0) h_t^n. \tag{4.2}$$

(4.2) gives the definition of $f(Z(t))$.

5. Itô Formula for Complex Brownian Motion

It follows immediately from the identity (4.2) that $f(Z(t))$ is a generalized complex Brownian functional. We state it as a theorem without proof.

Theorem 5.1: *Let $\psi \in \mathcal{SB}_\infty$. If $f \in \mathcal{SB}'_\infty(\mathbb{R})$, then $f(Z(t))$, defined by (4.2), is a member of $\mathcal{SB}'_\infty$. More precisely, for each $t > 0$, $\exists p \ni |h_t|_{-p} \leq 1$, and*

$$|\langle\!\langle f(Z(t)), \psi \rangle\!\rangle_c| \leq |f|_{-p} |\psi|_p.$$

Let $f \in \mathcal{SB}'_\infty(\mathbb{R})$. Then we have

$$\frac{d}{dt} \langle\!\langle f(Z(t)), \phi \rangle\!\rangle_c$$

$$= \sum_{n=0}^{\infty} b_n D^n \phi(0) h_t^{n-1} \delta_t = \sum_{n=0}^{\infty} b_n n D^{n-1}(D\phi(0)\delta_t) h_t^{n-1}$$

$$= \sum_{n=0}^{\infty} b_n n D^{n-1}(\partial_t \phi)(0) h_t^{n-1} = \langle\!\langle \partial_t^* f'(Z(t)), \phi \rangle\!\rangle_c$$

where $\partial_t = \partial_{\delta_t}$ and ∂_t^* is the adjoint operator of ∂_t. It follows that

$$\frac{d}{dt} f(Z(t)) = \partial_t^* f'(Z(t)).$$

This proves the Itô formula for complex Brownian motion. As a summary, we state the above result as a theorem.

Theorem 5.2: *Let $f \in \mathcal{SB}'_\infty(\mathbb{R})$. Then we have*

$$\frac{d}{dt} f(Z(t)) = \partial_t^* f'(Z(t))$$

or in the integral form,

$$f(Z(b)) - f(Z(a)) = \int_a^b \partial_t^* f'(Z(t)) dt.$$

As in the case of real Brownian motion, the term on the right hand side of Itô formula may be interpreted as stochastic integral as shown below.

Definition 5.3: Suppose that $f \in \mathcal{SB}'_\infty$. Define the stochastic integral $f(Z(t))$ as follows:

$$\left\langle\!\!\left\langle \int_a^b f(Z(t)) dZ(t), \phi \right\rangle\!\!\right\rangle_c$$

$$:= \lim_{\|\Delta_n\| \to 0} \left\langle\!\!\left\langle \sum_{i=1}^{n} f(Z(t_{i-1}))(Z(t_i) - Z(t_{i-1})), \phi \right\rangle\!\!\right\rangle_c$$

where $a = t_0 < t_1 < t_2 < \cdots < t_n = b$ and $\|\triangle_n\| = max_j|t_j - t_{j-1}|$.

Theorem 5.4: *Let $f \in \mathcal{SB}_\alpha(\mathbb{R})$ and $\phi \in \mathcal{SB}_\alpha$. Then*

$$\langle\!\langle \int_a^b f(Z(t))dZ(t), \phi \rangle\!\rangle_c = \langle\!\langle \int_a^b \partial_t^* f(Z(t))dt, \phi \rangle\!\rangle_c.$$

–A Connection between the Itô Formulas for the Complex and Real Brownian Motion

Recall the Itô formula of $f(t, B_t)$,

$$f(b, B_b) - f(a, B_a) = \int_a^b f_t(t, B_t)\, ds + \int_a^b f_x(t, B_t)dB_t$$
$$+ \frac{1}{2}\int_a^b f_{xx}(t, B_t)dt. \qquad (5.1)$$

Take S-transform, we obtain

$$\mu_b f(\langle\xi, h_b\rangle) - \mu_a f(\langle\xi, h_a\rangle) = \int_a^b \xi(t)(\mu_t f)'(\langle\xi, h_t\rangle)dt + \frac{1}{2}\mu_t f''(\langle\xi, h_t\rangle)dt, \qquad (5.2)$$

where $\mu_t f(u) = \int_\mathbb{R} f(u + \sqrt{t}v)\mu(dv)$.

Replace ξ by $Z(t)$ in the above equation, we have

$$\mu_b f(Z_b) - \mu_a(Z_a) = \int_a^b \mu_t f'(Z_t)dZ_t + \frac{1}{2}\int_a^b \mu f''(Z_t)dt. \qquad (5.3)$$

The above formula indeed follows from the Itô formula of complex Brownian motion by applying the Itô formula to $\mu_t f(t, Z_t)$:

$$\mu_b f(Z_b) - \mu_a(Z_a) = \int_a^b \mu_t f'(Z_t)dZ_t + \int_a^b \frac{d}{dt}(\mu_t f)(Z_t)dt, \qquad (5.4)$$

where the last term is verified by the following computation

$$\int_a^b \frac{d}{dt}(\mu_t f)(Z_t)dt$$
$$= \frac{1}{2}\int_a^b \frac{1}{\sqrt{t}}\int_\mathbb{R}[f'(Z_t + \sqrt{t}u)]\cdot u\,\mu(du)dt$$
$$= \frac{1}{2}\int_a^b \mu_t f''(Z_t)dt. \qquad (5.5)$$

Reverse the procedure, we can easily prove Itô formula for real Brownian motion from that of complex Brownian motion. In fact, start from the

Brownian motion $f(B(t)) = f(\langle x, h_t \rangle)$. Apply the Itô formula for the complex Brownian functional $\mu_t f(Z_t)$, we obtain the identity (5.4). Using the identity (5.5) one derive (5.3) from (5.4). Finally, replacing $\dot{Z}(t)$ by ξ, one prove that the identity (5.2). Then we prove the Itô formula by taking inverse S-transform.

As an example, we have

Example 5.5: To evaluate the integral

$$I = \int_a^b \partial_t^* B(1) \, dt.$$

We first take S-transform of I to obtain

$$S(I)(\xi) = e^{-\frac{1}{2}\|\xi\|_0^2} \int_a^b \langle\!\langle \partial_t^* B(1), e^{\langle \cdot, \xi \rangle} \rangle\!\rangle \, dt$$

$$= \int_a^b \xi(t) \langle \xi, h_1 \rangle \, dt.$$

Replace ξ by $\dot{Z}$, we obtain

$$S(I)(\dot{Z}) = \int_a^b \dot{Z}(t) \langle \dot{Z}, h_1 \rangle \, dt = (Z(b) - Z(a)) Z(1).$$

It follows that

$$I = \int_{\mathcal{S}'} \langle x + iy, h_b - h_a \rangle \langle x + iy, h_1 \rangle \, \mu(dy) = B(1)(B(b) - B(a)) - (b - a).$$

Remark 5.6: The above theory remains true that if we replace the Brownian motion by any non-adapted gaussian process. Thus the Itô formula for generalized fractional Brownian motions can be obtained from the above method. Since the computation is more involved, we shall discussed in another paper. The idea depends very much on the existence of S-transform, thus the method introduced in this chapter can also be applied to Lévy processes, for details, we refer the reader to [10, 11].

References

1. Ayed, W. and Kuo, H.H., An extension of the Itô formula, v. 2, *COSA* (2008) 323-333.
2. Hida, T., Kuo, H.-H., Potthoff, J., and Streit, L., *White Noise: An Infinite Dimensional Calculus*, Kluwer Academic Publishers, 1993.
3. Ito, Y. and Kubo, I., Calculus on Gaussian and Poisson White Noises, *Nagoya Math. J.* 111 (1988) 41-84.

4. Kubo, I. and Takenaka, S., Calculus on Gaussian White Noises I, *Proc. Japan Acad. Ser. A Math. Sci.* 56 (1980) 376-380; Calculus on Gaussian White Noises II, *Proc. Japan Acad. Ser. A Math. Sci.* 56 (1980) 411-416; Calculus on Gaussian White Noises III, *Proc. Japan Acad. Ser. A Math. Sci.* 57 (1981) 433-437; Calculus on Gaussian White Noise IV, *Proc. Japan Acad. Ser. A Math. Sci.* 58 (1982) 186-189.

5. Kuo, H.-H., *White Noise Distribution Theory*, CRC Press, 1996.

6. Lee, Y.-J., Generalized Functions on Infinite Dimensional Spaces and its Application to White Noise Calculus, *J. Funct. Anal.* 82 (1989) 429-464.

7. Lee, Y.-J., Analytic Version of Test Functionals, Fourier Transform and a Characterization of Measures in White Noise Calculus, *J. Funct. Anal.* 100 (1991) 359-380.

8. Lee, Y.-J., Integral Representation of Second Quantization and its Application to White Noise Analysis, *J. Funct. Anal.* 133 (1995) 253-276.

9. Lee, Y.-J., A characterization of generalized functions on infinite dimensional spaces and Bargmann-Segal Analytic functions, In *Gaussian Random Fields*, Series on Probability and Statistics Vol. 1 (1991) 272-284.

10. Lee, Y.-J. and Shih, H.-H., The Segal-Bargmann Transform for Lévy Functionals, *J. Funct. Anal.* 168 (1999) 46-83.

11. Lee, Y.-J. and Shih, H.-H., Analysis of generalized Lévy white noise functionals, *J. Funct. Anal.* 211 (2004) 1-70.

12. Lee, Y.-J. and Yen, K.-G., Analysis of compledx Brownian motion, *Commication of Stochastic Analysis*, Vol. 2, No. 1 (2008) 97-107.

13. Potthoff, J. and Streit, L., A Characterization of Hida Distributions, *J. Funct. Anal.* 101, 212-229.

QUASI QUANTUM QUADRATIC OPERATORS OF $\mathbb{M}_2(\mathbb{C})$

Farrukh Mukhamedov

Department of Mathematical Sciences
College of Science, United Arab Emirates University
P.O. Box 15551, Al Ain, Abu Dhabi, UAE
farrukh.m@uaeu.ac.ae
far75m@yandex.com

In the present chapter, we review on quasi quantum quadratic operators (q.q.o) acting on the algebra of 2×2 matrices $\mathbb{M}_2(\mathbb{C})$. Moreover, we provide new results on characterization of the q-purity of quasi q.q.o. This allowed us to describe all possibilities of quadratic operators associated with q-pure quasi q.q.o.

Keywords: Quasi quantum quadratic operators.

1. Introduction

In quantum information theory one of most important problem is the discrimination between separable and entangled states [16]. The most general tool consists in applying the theory of linear positive maps [19]. In this theory, it is essential to describe extremal elements of the set of unital linear positive mappings from algebra A to algebra B. In case $A = B = \mathbb{M}_2(\mathbb{C})$ the set of extremal elements is described in [23]. There it was showed that for each extreme point ϕ of the convex set of unital maps of $\mathbb{M}_2(\mathbb{C})$ into itself, there is a pure state ϕ of $\mathbb{M}_2(\mathbb{C})$ such that $\rho \circ \phi$ is a pure state. Therefore, it is natural to study maps $\phi : A \to B$, with A, B C^*-algebras, such that $\rho \circ \phi$ is a pure state for all pure states ρ of B. Such maps are called pure maps or pure channels (see [1]). For example, important examples of such kind of maps are conjugation of automorphisms of given algebra. But, if a channel acts from algebra to another one, then the description of pure channels is a tricky job. In [22] description of pure maps is given in case of $A = B(K)$, $B = B(H)$ where K, H are Hilbert spaces. Therefore, it would be interesting characterize such kind of maps (or channels). Note that quantum mutual entropy of such kind of maps can be calculated easier way than others [17, 18].

In [13] we have introduced a weaker condition then pure map. This notion is based on quadratic operators acting on the state space of the algebra. Note that quadratic operators are defined by quantum quadratic operators (see [4, 5, 10]), which are quantum generalization of well known quadratic systems [2, 7]. In all these investigations, the said quantum quadratic operators by definition are positive. But, in general, to study the nonlinear dynamics the positivity of the operator is strong condition. Therefore, in [13] we have introduced a weaker than the positivity, and corresponding operators are called *quasi quantum quadratic*. Each such kind of operator defines a quadratic operator acting on state space of $\mathbb{M}_2(\mathbb{C})$. In [13] we introduced a weaker condition, called q-purity, than purity of the mapping. To study q-pure channels, we concentrated ourselves to quasi q.q.o. acting on $\mathbb{M}_2(\mathbb{C})$.

In this chapter we review recent development on quasi quantum quadratic operators defined on $\mathbb{M}_2(\mathbb{C})$. Moreover, we provide new results on characterization of the q-purity of quasi q.q.o. This allowed us to describe all possibilities of quadratic operators associated with q-pure quasi q.q.o.

2. Preliminaries

Let $B(H)$ be the set of linear bounded operators from a complex Hilbert space H to itself. By $B(H) \otimes B(H)$ we mean tensor product of $B(H)$ into itself. In the sequel $\mathbb{1}$ means an identity matrix. By $B(H)^*$ it is usually denoted the conjugate space of $B(H)$. We recall that a linear functional $\varphi \in B(H)^*$ is called *positive* if $\varphi(x) \geq 0$ whenever $x \geq 0$. The set of all positive linear functionals is denoted by $B(H)^*_+$. A positive functional φ is called *state* if $\varphi(\mathbb{1}) = 1$. By $S(B(H))$ we denote the set of all states defined on $B(H)$.

Let $\Delta : B(H) \to B(H) \otimes B(H)$ be a linear operator. Then Δ defines a conjugate operator $\Delta^* : (B(H) \otimes B(H))^* \to B(H)^*$ by

$$\Delta^*(f)(x) = f(\Delta x), \ f \in (B(H) \otimes B(H))^*, \ x \in B(H).$$

One can define an operator V_Δ by

$$V_\Delta(\varphi) = \Delta^*(\varphi \otimes \varphi), \ \varphi \in B(H)^*.$$

Let $U : B(H) \otimes B(H) \to B(H) \otimes B(H)$ be a linear operator such that $U(x \otimes y) = y \otimes x$ for all $x, y \in \mathbb{M}_2(\mathbb{C})$.

Definition 2.1: A linear operator $\Delta : B(H) \to B(H) \otimes B(H)$ is said to be

(a) – a *quasi quantum quadratic operator (quasi q.q.o)* if it is unital (i.e. $\Delta \mathbf{1} = \mathbf{1} \otimes \mathbf{1}$), *-preserving (i.e. $\Delta(x^*) = \Delta(x)^*$, $\forall x \in B(H)$) and

$$V_\Delta(\varphi) \in B(H)_+^* \quad \text{whenever } \varphi \in B(H)_+^*;$$

(b) – a *quantum quadratic operator (q.q.o.)* if it is unital (i.e. $\Delta \mathbf{1} = \mathbf{1} \otimes \mathbf{1}$) and positive (i.e. $\Delta x \geq 0$ whenever $x \geq 0$);

(c) – a *symmetric* if one has $U\Delta = \Delta$.

One can see that if Δ is q.q.o. then it is a quasi q.q.o. A state $h \in S(B(H))$ is called *a Haar state* for a quasi q.q.o. Δ if for every $x \in B(H)$ one has

$$(h \otimes id) \circ \Delta(x) = (id \otimes h) \circ \Delta(x) = h(x)\mathbf{1}. \tag{1}$$

Remark 2.1: In [9] it has been studied symmetric q.q.o., which was called *quantum quadratic stochastic operator.*

Remark 2.2: We note that there is another approach to nonlinear quantum operators on C^*-algebras (see [8]).

Note that from unitality of Δ we conclude that for any quasi q.q.o. V_Δ maps $S(B(H))$ into itself. In some literature operator V_Δ is called *quadratic convolution* (see for example [6]). In [15] certain dynamical properties of V_Δ associated with q.q.o. defined on $\mathbb{M}_2(\mathbb{C})$ are investigated. In [11, 14] Kadison-Schwarz property of q.q.o. has been studied.

In quantum information, pure channels play important role, which can be defined as follows: a channel (i.e. positive and unital mapping) $T : B(H_1) \to B(H_2)$ is called *pure* if for any pure state $\varphi \in S(B(H_1))$ the state $T^*\varphi$ is also pure (see [1]). In [13] we have defined a more weaker notion of purity for quasi q.q.o. Namely, a quasi q.q.o. Δ is called *q-pure* if for any pure state φ the state $V_\Delta(\varphi)$ is also pure.

From this definition one can immediately see that purity of quasi q.q.o. implies its q-purity.

3. Quasi Quantum Quadratic Operators on $M_2(\mathbb{C})$

By $M_2(\mathbb{C})$ be an algebra of 2×2 matrices over complex field $\mathbb{C}$. In this section we are going to describe quantum quadratic operators on $M_2(\mathbb{C})$ as well as find necessary conditions for such operators to satisfy the Kadison-Schwarz property.

Recall [3] that the identity and Pauli matrices $\{\mathbf{1}, \sigma_1, \sigma_2, \sigma_3\}$ form a basis for $M_2(\mathbb{C})$, where

$$\sigma_1 = \begin{pmatrix} 0 & 1 \\ 1 & 0 \end{pmatrix} \quad \sigma_2 = \begin{pmatrix} 0 & -i \\ i & 0 \end{pmatrix} \quad \sigma_3 = \begin{pmatrix} 1 & 0 \\ 0 & -1 \end{pmatrix}.$$

In this basis every matrix $x \in M_2(\mathbb{C})$ can be written as $x = w_0 \mathbf{1} + \mathbf{w}\sigma$ with $w_0 \in \mathbb{C}$, $\mathbf{w} = (w_1, w_2, w_3) \in \mathbb{C}^3$, here $\mathbf{w}\sigma = w_1\sigma_1 + w_2\sigma_2 + w_3\sigma_3$. In what follows, we frequently use notation $\overline{\mathbf{w}} = (\overline{w_1}, \overline{w_2}, \overline{w_3})$.

Lemma 3.1: [20] *The following assertions hold true:*

(a) x is self-adjoint iff $w_0, \mathbf{w}$ are reals;
(b) $\mathrm{Tr}(x) = 1$ iff $w_0 = 0.5$, here Tr is the trace of a matrix x;
(c) $x > 0$ iff $\|\mathbf{w}\| \le w_0$, where $\|\mathbf{w}\| = \sqrt{|w_1|^2 + |w_2|^2 + |w_3|^2}$;
(d) A linear functional φ on $M_2(\mathbb{C})$ is a state iff it can be represented by

$$\varphi(w_0 \mathbf{1} + \mathbf{w}\sigma) = w_0 + \langle \mathbf{w}, \mathbf{f} \rangle, \qquad (2)$$

where $\mathbf{f} = (f_1, f_2, f_3) \in \mathbb{R}^3$ such that $\|\mathbf{f}\| \le 1$. Here as before $\langle \cdot, \cdot \rangle$ stands for the scalar product in $\mathbb{C}^3$.
(e) A state φ is a pure if and only if $\|\mathbf{f}\| = 1$. So pure states can be seen as the elements of unit sphere in $\mathbb{R}^3$.

Lemma 3.2: [13] *Let $x = w_0 \mathbf{1} \otimes \mathbf{1} + \mathbf{w}\sigma \otimes \mathbf{1} + \mathbf{1} \otimes \mathbf{r}\sigma$. Then the following statements hold true:*

(i) x is self-adjoint if and only if $w_0 \in \mathbb{R}$ and $\mathbf{w}, \mathbf{r} \in \mathbb{R}^3$;
(ii) x is positive if and only if $w_0 > 0$ and $\|\mathbf{w}\| + \|\mathbf{r}\| \le w_0$.

In the sequel we shall identify a state with a vector $\mathbf{f} \in \mathbb{R}^3$. By τ we denote a normalized trace, i.e. $\tau(x) = \frac{1}{2}\mathrm{Tr}(x)$, $x \in M_2(\mathbb{C})$.

Let $\Delta : M_2(\mathbb{C}) \to M_2(\mathbb{C}) \otimes M_2(\mathbb{C})$ be a quasi q.q.o. Then we write the operator Δ in terms of a basis in $M_2(\mathbb{C}) \otimes M_2(\mathbb{C})$ formed by the Pauli matrices. Namely,

$$\Delta \mathbf{1} = \mathbf{1} \otimes \mathbf{1};$$

$$\Delta(\sigma_i) = b_i(\mathbf{1} \otimes \mathbf{1}) + \sum_{j=1}^{3} b_{ji}^{(1)}(\mathbf{1} \otimes \sigma_j) + \sum_{j=1}^{3} b_{ji}^{(2)}(\sigma_j \otimes \mathbf{1})$$

$$+ \sum_{m,l=1}^{3} b_{ml,i}(\sigma_m \otimes \sigma_l), \quad i = 1, 2, 3. \tag{3}$$

In general, a description of positive operators is one of the main problems of quantum information. In the literature most tractable maps are positive and trace-preserving ones, since such maps arise naturally in quantum information theory (see [16]). Therefore, in the sequel we shall restrict ourselves to trace-preserving quasi q.q.o., i.e. $\tau \otimes \tau \circ \Delta = \tau$.

Proposition 3.1: [13] *Let* $\Delta : \mathbb{M}_2(\mathbb{C}) \to \mathbb{M}_2(\mathbb{C}) \otimes \mathbb{M}_2(\mathbb{C})$ *be a trace-preserving quasi q.q.o., then in* (3) *one has* $b_j = 0$, *and* $b_{ij}^{(1)}$, $b_{ij}^{(2)}$, $b_{ij,k}$ *are real for every* $i, j, k \in \{1, 2, 3\}$. *Moreover,* Δ *has the following form:*

$$\Delta(x) = w_0 \mathbf{1} \otimes \mathbf{1} + \mathbf{1} \otimes \mathbf{B}^{(1)} \mathbf{w} \cdot \sigma + \mathbf{B}^{(2)} \mathbf{w} \cdot \sigma \otimes \mathbf{1} + \sum_{m,l=1}^{3} \langle \mathbf{b}_{ml}, \overline{\mathbf{w}} \rangle \sigma_m \otimes \sigma_l, \tag{4}$$

where $x = w_0 \mathbf{1} + \mathbf{w}\sigma$, $\mathbf{b}_{ml} = (b_{ml,1}, b_{ml,2}, b_{ml,3})$, *and* $\mathbf{B}^{(k)} = (b_{ij}^{(k)})_{i,j=1}^{3}$, $k = 1, 2$. *Here as before* $\langle \cdot, \cdot \rangle$ *stands for the standard scalar product in* $\mathbb{C}^3$.

One can rewrite (4) as follows

$$\Delta(x) = \lambda \Delta_1(x) + (1 - \lambda)\Delta_2(x), \tag{5}$$

where

$$\Delta_1(x) = w_0 \mathbf{1} \otimes \mathbf{1} + \frac{1}{\lambda} \sum_{m,l=1}^{3} \langle \mathbf{b}_{ml}, \overline{\mathbf{w}} \rangle \sigma_m \otimes \sigma_l, \tag{6}$$

$$\Delta_2(x) = w_0 \mathbf{1} \otimes \mathbf{1} + \frac{1}{1-\lambda} \left(\mathbf{B}^{(2)} \mathbf{w} \cdot \sigma \otimes \mathbf{1} + \mathbf{1} \otimes \mathbf{B}^{(1)} \mathbf{w} \cdot \sigma \right). \tag{7}$$

Now assume that $b_{ij,k} = 0$ for all $i, j, k \in \{1, 2, 3\}$ and Δ is q-pure symmetric quasi q.q.o. In this case, Δ has the following form

$$\Delta(w_0 \mathbf{1} + \mathbf{w}\sigma) = w_0 \mathbf{1} \otimes \mathbf{1} + \mathbf{B}\mathbf{w} \cdot \sigma \otimes \mathbf{1} + \mathbf{1} \otimes \mathbf{B}\mathbf{w} \cdot \sigma. \tag{8}$$

Let us take any $\varphi \in S(\mathbb{M}_2(\mathbb{C}))$ and $\mathbf{f} \in \mathbb{R}^3$ be the corresponding vector. Then we find

$$\varphi \otimes \varphi(\Delta(w_0\mathbf{1} + \mathbf{w}\sigma)) = w_0 + 2\langle \mathbf{Bw}, \mathbf{f} \rangle = w_0 + \langle \mathbf{w}, 2\mathbf{B}^*\mathbf{f} \rangle.$$

Hence, if φ is pure, then $\|\mathbf{f}\| = 1$. Denoting $\mathbf{U} = 2\mathbf{B}^*$ and the q-purity of Δ yields that $\|\mathbf{Uf}\| = 1$ for all $\mathbf{f}$ with $\|\mathbf{f}\| = 1$. This means that $\mathbf{U}$ is isometry, so $\|\mathbf{U}\| = 1$, i.e. $\|\mathbf{B}\| = 1/2$. Consequently, one concludes that Δ is q-pure if and only if $2\mathbf{B}$ is isometry.

Now we are interested, whether q-pure symmetric quasi q.q.o. is positive. From Lemma 3.2 one gets the following

Theorem 3.1: *Let Δ be given by* (8). *Then the following statements hold true:*

(i) Δ is quasi q.q.o. if and only if Δ is positive, i.e. $\|\mathbf{B}\| \leq 1/2$;
(ii) Δ is q-pure if and only if $2\mathbf{B}$ is isometry. Moreover, Δ is positive.

Note that using the methods of [12] one may study Kadison-Schwarz property of mappings given by (8). Now the question is what about the case when $b_{ij,k} \neq 0$. Therefore, the next section is devoted to this this question.

4. q-Pure Symmetric Quasi Quantum Quadratic Operators on $\mathbb{M}_2(\mathbb{C})$

In this section, we are interested in the case when $\Delta_2 = 0$ in (5). This means that Δ has a Haar state τ. Then due to Lemma 3.1 (d) and Proposition 3.1 the functional $\Delta^*(\varphi \otimes \psi)$ is a state if and only if the vector

$$f_{\Delta^*(\varphi,\psi)} = \left(\sum_{i,j=1}^{3} b_{ij,1}f_ip_j, \sum_{i,j=1}^{3} b_{ij,2}f_ip_j, \sum_{i,j=1}^{3} b_{ij,3}f_ip_j \right) \qquad (9)$$

satisfies $\|f_{\Delta^*(\varphi,\psi)}\| \leq 1$.

Let us denote

$$\mathbf{B} = \{\mathbf{p} = (p_1, p_2, p_3) \in \mathbb{R}^3 : p_1^2 + p_2^2 + p_3^2 \leq 1\},$$
$$\mathbf{S} = \{\mathbf{p} = (p_1, p_2, p_3) \in \mathbb{R}^3 : p_1^2 + p_2^2 + p_3^2 = 1\},$$
$$\mathbf{C}_k = \{\mathbf{p} = \big((1 - \delta_{1k})p_1, (1 - \delta_{2k})p_2, (1 - \delta_{3k})p_3\big) \in \mathbf{S}^3\}, \quad k = 1, 2, 3,$$

where δ_{ij} is the usual Kronecker delta.

Let us consider the quadratic operator defined by $V_\Delta(\varphi) = \Delta^*(\varphi \otimes \varphi)$, $\varphi \in S(\mathbb{M}_2(\mathbb{C}))$. From (9) we find that

$$V_\Delta(\varphi)(\sigma_k) = \sum_{i,j=1}^{3} b_{ij,k} f_i f_j, \quad \mathbf{f} \in \mathbf{B}.$$

This suggests us the consideration of a nonlinear operator $V : \mathbf{B} \to \mathbf{B}$ defined by

$$V(\mathbf{f})_k = \sum_{i,j=1}^{3} b_{ij,k} f_i f_j, \quad k = 1,2,3, \tag{10}$$

where $\mathbf{f} = (f_1, f_2, f_3) \in \mathbf{B}$.

From the definition and Lemma 3.1 (e) we conclude that the Δ is q-pure if and only if $V(\mathbf{S}) \subset \mathbf{S}$.

Now let us rewrite the quadratic operator V (see (10)) as follows

$$V(\mathbf{f}) = \begin{cases} a_1 f_1^2 + b_1 f_2^2 + c_1 f_3^2 + \alpha_1 f_1 f_2 + \beta_1 f_2 f_3 + \gamma_1 f_1 f_3 \\ a_2 f_1^2 + b_2 f_2^2 + c_2 f_3^2 + \alpha_2 f_1 f_2 + \beta_2 f_2 f_3 + \gamma_2 f_1 f_3 \\ a_3 f_1^2 + b_3 f_2^2 + c_3 f_3^2 + \alpha_3 f_1 f_2 + \beta_3 f_2 f_3 + \gamma_3 f_1 f_3 \end{cases} \tag{11}$$

where $\mathbf{f} \in \mathbf{B}$.

Theorem 4.1: [13] *Let the operator V given by* (11). *Then $V(\mathbf{S}) \subset \mathbf{S}$ if and only if the followings hold true*

(i) $\|\mathbf{a}\| = 1$, $\|\mathbf{b}\| = 1$, $\|\mathbf{c}\| = 1$;
(ii) $\|A\| = \|\mathbf{a} - \mathbf{b}\|$, $\|\Gamma\| = \|\mathbf{a} - \mathbf{c}\|$, $\|B\| = \|\mathbf{b} - \mathbf{c}\|$;
(iii) $\langle \mathbf{a}, B \rangle + \langle A, \Gamma \rangle = 0$, $\langle \mathbf{b}, \Gamma \rangle + \langle A, B \rangle = 0$, $\langle \mathbf{c}, A \rangle + \langle B, \Gamma \rangle = 0$;
(iv) $\langle \mathbf{a}, A \rangle = 0$, $\langle \mathbf{a}, \Gamma \rangle = 0$, $\langle \mathbf{b}, A \rangle = 0$, $\langle \mathbf{b}, B \rangle = 0$, $\langle \mathbf{c}, \Gamma \rangle = 0$, $\langle \mathbf{c}, B \rangle = 0$

where $\mathbf{a} = (a_1, a_2, a_3)$, $\mathbf{b} = (b_1, b_2, b_3)$, $\mathbf{c} = (c_1, c_2, c_3)$, $\Gamma = (\gamma_1, \gamma_2, \gamma_3)$, $A = (\alpha_1, \alpha_2, \alpha_3)$, $B = (\beta_1, \beta_2, \beta_3)$.

Using the same argument one can prove the following

Theorem 4.2: *Let V be given by* (11). *Then V maps $\mathbf{S}$ to $\mathbf{C}_k$ ($k = 1, 2, 3$) if and only if the followings conditions hold:*

(i) $\|\mathbf{a}\| = 1$, $\|\mathbf{b}\| = 1$, $\|\mathbf{c}\| = 1$;
(ii) $\|A\| = \|\mathbf{a} - \mathbf{b}\|$, $\|\Gamma\| = \|\mathbf{a} - \mathbf{c}\|$, $\|B\| = \|\mathbf{b} - \mathbf{c}\|$;
(iii) $\langle \mathbf{a}, B \rangle + \langle A, \Gamma \rangle = 0$, $\langle \mathbf{b}, \Gamma \rangle + \langle A, B \rangle = 0$, $\langle \mathbf{c}, A \rangle + \langle B, \Gamma \rangle = 0$;
(iv) $\langle \mathbf{a}, A \rangle = 0$, $\langle \mathbf{a}, \Gamma \rangle = 0$, $\langle \mathbf{b}, A \rangle = 0$, $\langle \mathbf{b}, B \rangle = 0$, $\langle \mathbf{c}, \Gamma \rangle = 0$, $\langle \mathbf{c}, B \rangle = 0$,

where $\mathbf{a} = ((1 - \delta_{1k})a_1, (1 - \delta_{2k})a_2, (1 - \delta_{3k})a_3)$, $\mathbf{b} = ((1 - \delta_{1k})b_1, (1 - \delta_{2k})b_2, (1 - \delta_{3k})b_3)$, $\mathbf{c} = ((1 - \delta_{1k})c_1, (1 - \delta_{2k})c_2, (1 - \delta_{3k})c_3)$, $\Gamma = ((1 - \delta_{1k})\gamma_1, (1 - \delta_{2k})\gamma_2, (1 - \delta_{3k})\gamma_3)$, $A = ((1 - \delta_{1k})\alpha_1, (1 - \delta_{2k})\alpha_2, (1 - \delta_{3k})\alpha_3)$, $B = ((1 - \delta_{1k})\beta_1, (1 - \delta_{2k})\beta_2, (1 - \delta_{3k})\beta_3)$.

Now let us investigate a symmetric quasi q.q.o. Δ corresponding to (11). According to Proposition 3.1 the operator Δ has the following form

$$\Delta(x) = \begin{pmatrix} w_0 + \langle \mathbf{c}, \mathbf{w} \rangle & \frac{1}{2}s(\Gamma, -iB, \mathbf{w}) & \frac{1}{2}s(\Gamma, -iB, \mathbf{w}) & s(\mathbf{a} - \mathbf{b}, -iA, \mathbf{w}) \\ \frac{1}{2}s(\Gamma, iB, \mathbf{w}) & w_0 - \langle \mathbf{c}, \mathbf{w} \rangle & s(\mathbf{a}, \mathbf{b}, \mathbf{w}) & -\frac{1}{2}s(\Gamma, -iB, \mathbf{w}) \\ \frac{1}{2}s(\Gamma, iB, \mathbf{w}) & s(\mathbf{a}, \mathbf{b}, \mathbf{w}) & w_0 - \langle \mathbf{c}, \mathbf{w} \rangle & -\frac{1}{2}s(\Gamma, -iB, \mathbf{w}) \\ s(\mathbf{a} - \mathbf{b}, iA, \mathbf{w}) & -\frac{1}{2}s(\Gamma, iB, \mathbf{w}) & -\frac{1}{2}s(\Gamma, iB, \mathbf{w}) & w_0 + \langle \mathbf{c}, \mathbf{w} \rangle \end{pmatrix}$$

where $s(\mathbf{x}, \mathbf{y}, \mathbf{z}) = \langle \mathbf{x} + \mathbf{y}, \mathbf{z} \rangle$.

Theorem 4.3: *Let $\Delta : \mathbb{M}_2(\mathbb{C}) \to \mathbb{M}_2(\mathbb{C}) \otimes \mathbb{M}_2(\mathbb{C})$ be a symmetric quasi q.q.o. with Haar state τ and corresponding quadratic operator V maps $\mathbf{S}$ to $\mathbf{C}_k$ ($k = 1, 2, 3$). Then Δ is not positive.*

Now we are going to investigate V which maps the unite circle into the sphere.

Theorem 4.4: *The operator V given by (11). Then $V(\mathbf{C}_3) \subset \mathbf{S}$ if and only if one of the followings hold:*

(i) $\mathbf{a} = \mathbf{b}$, $A = 0$, $\|\mathbf{a}\| = 1$;
(ii) $\mathbf{a} = -\mathbf{b}$, $\|A\| = 2$, $\langle A, \mathbf{a} \rangle = 0$ *and* $\|\mathbf{a}\| = 1$;
(iii) $\mathbf{a}$ *is not parallel to* $\mathbf{b}$, $A = \lambda[\mathbf{a}, \mathbf{b}]$, $\|A\| = \|\mathbf{a} - \mathbf{b}\|$ *and* $\|\mathbf{a}\| = \|\mathbf{b}\| = 1$.

Remark 4.1: By the same argument one can prove the similar results for C_1 and C_2, respectively.

Now we turn to the positivity of Δ which corresponds to the operator V given by (11).

Theorem 4.5: *Let V given by (11) and $V(\mathbf{C}_3) \subset \mathbf{S}$. Assume that (i) of Theorem 4.4 is satisfied. Then a symmetric quasi q.q.o. $\Delta : \mathbb{M}_2(\mathbb{C}) \to \mathbb{M}_2(\mathbb{C}) \otimes \mathbb{M}_2(\mathbb{C})$ corresponding to V is not positive.*

Proof: Now we are going to prove from contrary. Assume (i) of Theorem 4.4 is satisfied and Δ is positive. This implies that the matrix given by (12) should be positive, when x is positive.

Let $x = \mathbf{1} + \mathbf{a}\sigma$, then (12) has the following form

$$\Delta(x) = \begin{pmatrix} 1 + \langle \mathbf{c}, \mathbf{a} \rangle & \frac{1}{2}s(\Gamma, -iB, \mathbf{a}) & \frac{1}{2}s(\Gamma, -iB, \mathbf{a}) & 0 \\ \frac{1}{2}s(\Gamma, iB, \mathbf{a}) & 1 - \langle \mathbf{c}, \mathbf{a} \rangle & 2 & -\frac{1}{2}s(\Gamma, -iB, \mathbf{a}) \\ \frac{1}{2}s(\Gamma, iB, \mathbf{a}) & 2 & 1 - \langle \mathbf{c}, \mathbf{a} \rangle & -\frac{1}{2}s(\Gamma, -iB, \mathbf{a}) \\ 0 & -\frac{1}{2}s(\Gamma, iB, \mathbf{a}) & -\frac{1}{2}s(\Gamma, iB, \mathbf{a}) & 1 + \langle \mathbf{c}, \mathbf{a} \rangle \end{pmatrix}.$$

Then calculation show that two eigenvalues of the matrix $\Delta(x)$ are followings $\lambda_1 = 1 + \langle \mathbf{c}, \mathbf{a} \rangle$, $\lambda_2 = -1 - \langle \mathbf{c}, \mathbf{a} \rangle$. The positivity yields that $\lambda_1 \geq 0$, $\lambda_2 \geq 0$ therefore, one finds $\langle \mathbf{c}, \mathbf{a} \rangle = -1$.

Now let $x = \mathbf{1} - \mathbf{a}\sigma$, then we have

$$\Delta(x) = \begin{pmatrix} 1 - \langle \mathbf{c}, \mathbf{a} \rangle & -\frac{1}{2}s(\Gamma, -iB, \mathbf{a}) & -\frac{1}{2}s(\Gamma, -iB, \mathbf{a}) & 0 \\ -\frac{1}{2}s(\Gamma, iB, \mathbf{a}) & 1 + \langle \mathbf{c}, \mathbf{a} \rangle & -2 & \frac{1}{2}s(\Gamma, -iB, \mathbf{a}) \\ -\frac{1}{2}s(\Gamma, iB, \mathbf{a}) & -2 & 1 + \langle \mathbf{c}, \mathbf{a} \rangle & \frac{1}{2}s(\Gamma, -iB, \mathbf{a}) \\ 0 & \frac{1}{2}s(\Gamma, iB, \mathbf{a}) & \frac{1}{2}s(\Gamma, iB, \mathbf{a}) & 1 - \langle \mathbf{c}, \mathbf{a} \rangle \end{pmatrix}.$$

A simple algebra shows us that one eigenvalue of $\Delta(x)$ can be written as follows

$$\lambda = -\sqrt{1 + \langle \mathbf{c}, \mathbf{a} \rangle^2 + \langle B, \mathbf{a} \rangle^2 + \langle \Gamma, \mathbf{a} \rangle^2 - 2\langle \mathbf{c}, \mathbf{a} \rangle}.$$

Using $\langle \mathbf{c}, \mathbf{a} \rangle = -1$ one gets

$$\lambda = -\sqrt{4 + \langle B, \mathbf{a} \rangle^2 + \langle \Gamma, \mathbf{a} \rangle^2}.$$

This means that $\lambda < 0$, which contradicts to $\Delta(x) \geq 0$. This completes the proof. $\qquad\square$

Remark 4.2: We note that using the same argument as in the proof of Theorem 4.5 it can be showed that Δ corresponding to V with $V(\mathbf{C}_k) \subset \mathbf{S}$ $(k = 1, 2)$ is not positive.

We need the following auxiliary result.

Lemma 4.1: *If*

$$xyz(Ax + By + Cz) \leq 0 \quad \text{for all} \quad (x, y, z) \in \mathbf{S}, \tag{12}$$

then $A = B = C = 0$.

In general, to check q-purity of a given quasi q.q.o. is tricky. Therefore, we are interested in finding simple conditions to ensure the q-purity.

Theorem 4.6: *Let* $\Delta : \mathbb{M}_2(\mathbb{C}) \to \mathbb{M}_2(\mathbb{C}) \otimes \mathbb{M}_2(\mathbb{C})$ *be a quasi q.q.o. and assume the corresponding quadratic operator V is given by (11). Then Δ is q-pure if and only if $V(\mathbf{C}_k) \subset \mathbf{S}$ for all $k = 1, 2, 3$.*

Proof: "if" part is obvious. Therefore, we start with "only if" part. Now assume $V(\mathbf{C}_k) \subset \mathbf{S}$ is valid for all $k = 1, 2, 3$. Then due to Theorem 4.4 and Remark 4.1 one finds

(1.1) $\|\mathbf{b}\| = 1$, $\|\mathbf{c}\| = 1$;
(1.2) $\langle B, \mathbf{b}\rangle = 0$, $\langle B, \mathbf{c}\rangle = 0$;
(1.3) $\|B\| = \|\mathbf{b} - \mathbf{c}\|$;
(2.1) $\|\mathbf{a}\| = 1$, $\|\mathbf{c}\| = 1$;
(2.2) $\langle \Gamma, \mathbf{a}\rangle = 0$, $\langle \Gamma, \mathbf{c}\rangle = 0$;
(2.3) $\|\Gamma\| = \|\mathbf{a} - \mathbf{c}\|$;
(3.1) $\|\mathbf{a}\| = 1$, $\|\mathbf{b}\| = 1$;
(3.2) $\langle A, \mathbf{a}\rangle = 0$, $\langle A, \mathbf{b}\rangle = 0$;
(3.3) $\|\Gamma\| = \|\mathbf{a} - \mathbf{b}\|$;

Let $\mathbf{f} \in \mathbf{S}$, then from quasiness of Δ we have $V(\mathbf{f}) \in \mathbf{B}$. This means

$$\left(V(\mathbf{f})_1\right)^2 + \left(V(\mathbf{f})_2\right)^2 + \left(V(\mathbf{f})_3\right)^2 \leq 1.$$

From (11) the last inequality can be rewritten as follows

$$\left(\|\mathbf{a} - \mathbf{c}\|^2 - \|\Gamma\|^2\right)f_1^4 + \left(\|\mathbf{b} - \mathbf{c}\|^2 - \|B\|^2\right)f_2^4$$

$$+ 2\left(\langle \mathbf{a} - \mathbf{c}, A\rangle - \langle B, \Gamma\rangle\right)f_1^3 f_2$$

$$+ 2\langle \mathbf{a} - \mathbf{c}, \Gamma\rangle f_1^3 f_3 + 2\left(\langle \mathbf{b} - \mathbf{c}, A\rangle - \langle B, \Gamma\rangle\right)f_1 f_2^3 + 2\langle \mathbf{b} - \mathbf{c}, B\rangle f_2^3 f_3$$

$$+ \left(\|\mathbf{a} - \mathbf{c}\|^2 + \|\mathbf{b} - \mathbf{c}\|^2 - \|\mathbf{a} - \mathbf{b}\|^2 + \|A\|^2 - \|B\|^2 - \|\Gamma\|^2\right)f_1^2 f_2^2$$

$$+ 2\left(\langle \mathbf{a} - \mathbf{c}, B\rangle + \langle A, \Gamma\rangle\right)f_1^2 f_2 f_3 + \left(\|\mathbf{a}\|^2 - \|\mathbf{c}\|^2 - \|\mathbf{a} - \mathbf{c}\|^2 + \|\Gamma\|^2\right)f_1^2$$

$$+ 2\left(\langle \mathbf{b} - \mathbf{c}, \Gamma\rangle + \langle A, B\rangle\right)f_1 f_2^2 f_3 + \left(\|\mathbf{b}\|^2 - \|\mathbf{c}\|^2 - \|\mathbf{b} - \mathbf{c}\|^2 + \|B\|^2\right)f_2^2$$

$$+ 2\left(\langle \mathbf{c}, A\rangle + \langle B, \Gamma\rangle\right)f_1 f_2 + 2\langle \mathbf{c}, \Gamma\rangle f_1 f_3 + 2\langle \mathbf{c}, B\rangle f_2 f_3 + \|\mathbf{c}\|^2 \leq 1.$$

Taking into account (1.1)–(3.3) we obtain

$$f_1 f_2 \left(\left(\langle \mathbf{c}, A \rangle + \langle B, \Gamma \rangle \right) \left(1 - f_1^2 - f_2^2 \right) + \left(\langle \mathbf{a}, B \rangle + \langle A, \Gamma \rangle \right) f_1 f_3 \right.$$

$$\left. + \left(\langle \mathbf{b}, \Gamma \rangle + \langle A, B \rangle \right) f_2 f_3 \right) \leq 0.$$

The last inequality yields

$$f_1 f_2 f_3 \left(\left(\langle \mathbf{c}, A \rangle + \langle B, \Gamma \rangle \right) f_3 + \left(\langle \mathbf{a}, B \rangle + \langle A, \Gamma \rangle \right) f_1 + \left(\langle \mathbf{b}, \Gamma \rangle + \langle A, B \rangle \right) f_2 \right)$$

$$\leq 0.$$

Due to Lemma 4.1 one finds

$$\langle \mathbf{c}, A \rangle + \langle B, \Gamma \rangle = 0, \quad \langle \mathbf{a}, B \rangle + \langle A, \Gamma \rangle = 0, \quad \langle \mathbf{b}, \Gamma \rangle + \langle A, B \rangle = 0. \quad (13)$$

According to Theorem 4.1 and (1.1)–(3.3) with (13) yields $V(\mathbf{S}) \subset \mathbf{S}$. This completes the proof. □

Remark 4.3: We should stress that without quasiness of Δ Theorem 4.6 fails (see Corollary 4.1).

Remark 4.4: Note that q-purity condition of the operator is well defined. The last theorem implies that quadratic operator V maps pure state to pure ones iff it maps 3 unit circles to the sphere at the same time.

Now we want to describe quadratic operators corresponding to q-pure quasi q.q.o. Let V be given by (11) and it maps $\mathbf{C}_1, \mathbf{C}_2, \mathbf{C}_3$ to $\mathbf{S}$ at the same time.

Theorem 4.7: *Let V be given by (11). Then $V(\mathbf{C}_k) \subset \mathbf{S}$, $k = 1, 2, 3$ hold if and only if it has one of the following forms:*

(1)

$$V_1(\mathbf{f}) = \begin{cases} a_1 \left(f_1^2 + f_2^2 + f_3^2 \right) \\ a_2 \left(f_1^2 + f_2^2 + f_3^2 \right) \\ a_3 \left(f_1^2 + f_2^2 + f_3^2 \right) \end{cases}$$

    where $\|\mathbf{a}\| = 1$;

(2)

$$V_2(\mathbf{f}) = \begin{cases} a_1 \left(f_1^2 - f_2^2 - f_3^2 \right) + \alpha_1 f_1 f_2 + \gamma_1 f_1 f_3 \\ a_2 \left(f_1^2 - f_2^2 - f_3^2 \right) + \alpha_2 f_1 f_2 + \gamma_2 f_1 f_3 \\ a_3 \left(f_1^2 - f_2^2 - f_3^2 \right) + \alpha_3 f_1 f_2 + \gamma_3 f_1 f_3 \end{cases} ,$$

where $\|A\| = 2$, $\|\Gamma\| = 2$, $\langle A, \mathbf{a} \rangle = 0$, $\langle \Gamma, \mathbf{a} \rangle = 0$, *and* $\|\mathbf{a}\| = 1$;

(3)

$$V_3(\mathbf{f}) = \begin{cases} a_1 f_1^2 + b_1 \big(f_2^2 + f_3^2\big) + \alpha_1 f_1 f_2 + \gamma_1 f_1 f_3 \\ a_2 f_1^2 + b_2 \big(f_2^2 + f_3^2\big) + \alpha_2 f_1 f_2 + \gamma_2 f_1 f_3 \\ a_3 f_1^2 + b_3 \big(f_2^2 + f_3^2\big) + \alpha_3 f_1 f_2 + \gamma_3 f_1 f_3 \end{cases},$$

where $\mathbf{a} \not\parallel \mathbf{b}$, $A = \lambda[\mathbf{a}, \mathbf{b}]$, $\Gamma = \mu[\mathbf{a}, \mathbf{c}]$, $\|A\| = \|\mathbf{a} - \mathbf{b}\|$, $\|\Gamma\| = \|\mathbf{a} - \mathbf{c}\|$ *and* $\|\mathbf{a}\| = \|\mathbf{b}\| = 1$;

(4)

$$V_4(\mathbf{f}) = \begin{cases} a_1 \big(f_1^2 - f_2^2 + f_3^2\big) + \alpha_1 f_1 f_2 + \beta_1 f_2 f_3 \\ a_2 \big(f_1^2 - f_2^2 + f_3^2\big) + \alpha_2 f_1 f_2 + \beta_2 f_2 f_3 \\ a_3 \big(f_1^2 - f_2^2 + f_3^2\big) + \alpha_3 f_1 f_2 + \beta_3 f_2 f_3 \end{cases},$$

where $\|A\| = 2$, $\|B\| = 2$, $\langle A, \mathbf{a} \rangle = 0$, $\langle B, \mathbf{a} \rangle = 0$ *and* $\|\mathbf{a}\| = 1$;

(5)

$$V_5(\mathbf{f}) = \begin{cases} a_1 \big(f_1^2 + f_2^2 - f_3^2\big) + \beta_1 f_2 f_3 + \gamma_1 f_1 f_3 \\ a_2 \big(f_1^2 + f_2^2 - f_3^2\big) + \beta_2 f_2 f_3 + \gamma_2 f_1 f_3 \\ a_3 \big(f_1^2 + f_2^2 - f_3^2\big) + \beta_3 f_2 f_3 + \gamma_3 f_1 f_3 \end{cases},$$

where $\|B\| = 2$, $\|\Gamma\| = 2$, $\langle B, \mathbf{a} \rangle = \langle \Gamma, \mathbf{a} \rangle = 0$ *and* $\|\mathbf{a}\| = 1$;

(6)

$$V_6(\mathbf{f}) = \begin{cases} a_1 f_1^2 + b_1 \big(f_2^2 - f_3^2\big) + \alpha_1 f_1 f_2 + \beta_1 f_2 f_3 + \gamma_1 f_1 f_3 \\ a_2 f_1^2 + b_2 \big(f_2^2 - f_3^2\big) + \alpha_2 f_1 f_2 + \beta_2 f_2 f_3 + \gamma_2 f_1 f_3 \\ a_3 f_1^2 + b_3 \big(f_2^2 - f_3^2\big) + \alpha_3 f_1 f_2 + \beta_3 f_2 f_3 + \gamma_3 f_1 f_3 \end{cases},$$

where $\mathbf{a} \not\parallel \mathbf{b}$, $\|B\| = 2$, $A = \lambda[\mathbf{a}, \mathbf{b}]$, $\Gamma = \mu[\mathbf{a}, \mathbf{b}]$, $\|A\| = \|\mathbf{a} - \mathbf{b}\|$, $\|\Gamma\| = \|\mathbf{a} + \mathbf{b}\|$, $\langle B, \mathbf{b} \rangle = 0$, *and* $\|\mathbf{a}\| = \|\mathbf{b}\| = 1$;

(7)

$$V_7(\mathbf{f}) = \begin{cases} a_1 \big(f_1^2 + f_3^2\big) + b_1 f_2^2 + \alpha_1 f_1 f_2 + \beta_1 f_2 f_3 \\ a_2 \big(f_1^2 + f_3^2\big) + b_2 f_2^2 + \alpha_2 f_1 f_2 + \beta_2 f_2 f_3 \\ a_3 \big(f_1^2 + f_3^2\big) + b_3 f_2^2 + \alpha_3 f_1 f_2 + \beta_3 f_2 f_3 \end{cases},$$

where $\mathbf{a} \not\parallel \mathbf{b}$, $A = \lambda[\mathbf{a}, \mathbf{b}]$, $B = \mu[\mathbf{b}, \mathbf{a}]$, $\|A\| = \|B\| = \|\mathbf{a} - \mathbf{b}\|$ *and* $\|\mathbf{a}\| = \|\mathbf{b}\| = 1$;

(8)

$$V_8(\mathbf{f}) = \begin{cases} a_1 \big(f_1^2 - f_3^2\big) + b_1 f_2^2 + \alpha_1 f_1 f_2 + \beta_1 f_2 f_3 + \gamma_1 f_1 f_3 \\ a_2 \big(f_1^2 - f_3^2\big) + b_2 f_2^2 + \alpha_2 f_1 f_2 + \beta_2 f_2 f_3 + \gamma_2 f_1 f_3 \\ a_3 \big(f_1^2 - f_3^2\big) + b_3 f_2^2 + \alpha_3 f_1 f_2 + \beta_3 f_2 f_3 + \gamma_3 f_1 f_3 \end{cases},$$

where $\mathbf{a} \not\parallel \mathbf{b}$, $\|\Gamma\| = 2$, $A = \lambda[\mathbf{a}, \mathbf{b}]$, $B = \mu[\mathbf{a}, \mathbf{b}]$, $\|A\| = \|\mathbf{a} - \mathbf{b}\|$, $\|B\| = \|\mathbf{b} + \mathbf{a}\|$, $\langle \Gamma, \mathbf{a} \rangle = 0$ *and* $\|\mathbf{a}\| = \|\mathbf{b}\| = 1$;

(9)

$$V_9(\mathbf{f}) = \begin{cases} a_1\left(f_1^2 + f_2^2\right) + c_1 f_3^2 + \beta_1 f_2 f_3 + \gamma_1 f_1 f_3 \\ a_2\left(f_1^2 + f_2^2\right) + c_2 f_3^2 + \beta_2 f_2 f_3 + \gamma_2 f_1 f_3 \\ a_3\left(f_1^2 + f_2^2\right) + c_3 f_3^2 + \beta_3 f_2 f_3 + \gamma_3 f_1 f_3 \end{cases},$$

where $\mathbf{a} \parallel\!\!\!/ \mathbf{c}$, $B = \lambda[\mathbf{a}, \mathbf{c}]$, $\Gamma = \mu[\mathbf{a}, \mathbf{c}]$, $\|B\| = \|\Gamma\| = \|\mathbf{a} - \mathbf{c}\|$ *and* $\|\mathbf{a}\| = \|\mathbf{c}\| = 1$;

(10)

$$V_{10}(\mathbf{f}) = \begin{cases} a_1\left(f_1^2 - f_2^2\right) + c_1 f_3^2 + \alpha_1 f_1 f_2 + \beta_1 f_2 f_3 + \gamma_1 f_1 f_3 \\ a_2\left(f_1^2 - f_2^2\right) + c_2 f_3^2 + \alpha_2 f_1 f_2 + \beta_2 f_2 f_3 + \gamma_2 f_1 f_3 \\ a_3\left(f_1^2 - f_2^2\right) + c_3 f_3^2 + \alpha_3 f_1 f_2 + \beta_3 f_2 f_3 + \gamma_3 f_1 f_3 \end{cases},$$

where $\mathbf{a} \parallel\!\!\!/ \mathbf{c}$, $\|A\| = 2$, $B = \lambda[\mathbf{a}, \mathbf{c}]$, $\Gamma = \mu[\mathbf{a}, \mathbf{c}]$, $\|B\| = \|\mathbf{a} + \mathbf{c}\|$, $\|\Gamma\| = \|\mathbf{a} - \mathbf{c}\|$, $\langle A, \mathbf{a} \rangle = 0$ *and* $\|\mathbf{a}\| = \|\mathbf{c}\| = 1$;

(11)

$$V_{11}(\mathbf{f}) = \begin{cases} a_1 f_1^2 + b_1 f_2^2 + c_1 f_3^2 + \alpha_1 f_1 f_2 + \beta_1 f_2 f_3 + \gamma_1 f_1 f_3 \\ a_2 f_1^2 + b_2 f_2^2 + c_2 f_3^2 + \alpha_2 f_1 f_2 + \beta_2 f_2 f_3 + \gamma_2 f_1 f_3 \\ a_3 f_1^2 + b_3 f_2^2 + c_3 f_3^2 + \alpha_3 f_1 f_2 + \beta_3 f_2 f_3 + \gamma_3 f_1 f_3 \end{cases}$$

where $\mathbf{b} \parallel\!\!\!/ \mathbf{c}$, $\mathbf{a} \parallel\!\!\!/ \mathbf{c}$, $\mathbf{a} \parallel\!\!\!/ \mathbf{b}$, $A = \delta[\mathbf{a}, \mathbf{b}]$, $B = \lambda[\mathbf{b}, \mathbf{c}]$, $\Gamma = \mu[\mathbf{a}, \mathbf{c}]$, $\|A\| = \|\mathbf{a} - \mathbf{b}\|$, $\|B\| = \|\mathbf{b} - \mathbf{c}\|$, $\|\Gamma\| = \|\mathbf{a} - \mathbf{c}\|$ *and* $\|\mathbf{a}\| = \|\mathbf{b}\| = \|\mathbf{c}\| = 1$.

Proposition 4.1: *Let V be given by (11) and assume $V(\mathbf{C}_k) \subset \mathbf{S}$ ($k = 1, 2, 3$). Then the operator V belongs to one of the following classes (non isomorphic)*

$$K_1 = \{V_1\}, \quad K_2 = \{V_2, V_4, V_5\}, \quad K_3 = \{V_3, V_7, V_9\},$$
$$K_4 = \{V_6, V_8, V_{10}\}, \quad K_5 = \{V_{11}\}.$$

Theorem 4.8: *Let V be given by (11) and assume $V(\mathbf{C}_k) \subset \mathbf{S}$ ($k = 1, 2, 3$). Then $\Delta : \mathrm{M}_2(\mathbb{C}) \to \mathrm{M}_2(\mathbb{C}) \otimes \mathrm{M}_2(\mathbb{C})$ corresponding to V is quasi q.q.o. if and only if V belongs to one of the following classes:*

$$K_1 = \{V_1\}, \quad K_2 = \{V_2, V_4, V_5\},$$
$$K_4 = \{V_6, V_8, V_{10}\}, \quad K_5 = \{V_{11}\}.$$

Corollary 4.1: *Let V belong to K_3. Then the corresponding linear operator $\Delta : \mathrm{M}_2(\mathbb{C}) \to \mathrm{M}_2(\mathbb{C}) \otimes \mathrm{M}_2(\mathbb{C})$ is not quasi q.q.o. Namely, $V(\mathbf{S}) \not\subset \mathbf{B}$.*

Proof: Let $V \in K_3$. Without loss of generality we may assume $V = V_3$. According to (3) of Theorem 4.7 we conclude that $A \parallel \Gamma$, i.e. $\langle A, \Gamma \rangle \neq 0$.

Let $\mathbf{f} \in \mathbf{S}$ and show $V(\mathbf{f}) \notin \mathbf{B}$. Indeed, using (3) of Theorem 4.7 one gets

$$\|V(\mathbf{f})\|^2 = 1 + 2\langle A, \Gamma \rangle f_1^2 f_2 f_3. \tag{14}$$

Choose f_1, f_2, f_3 such a way that

$$Sign(f_2 f_3) = Sign(\langle A, \Gamma \rangle), \quad f_1 f_2 f_3 \neq 0.$$

Then from (14) one finds

$$1 + 2\langle A, \Gamma \rangle f_1^2 f_2 f_3 > 1.$$

This means $V(\mathbf{f}) \notin \mathbf{B}$. This completes the proof. $\qquad\qquad\square$

Remark 4.5: We note that in [21] a concrete example of not quasi q.q.o. is constructed, for which $V(\mathbf{C}_k) \subset \mathbf{S}$ ($k = 1, 2, 3$) holds. Our Corollary 4.1 exhibits all possibilities of such kind of operators.

5. Conclusion

In this chapter we studied quasi quantum quadratic operators (q.q.o) acting on the algebra of 2×2 matrices $\mathbb{M}_2(\mathbb{C})$. We considered two kinds of quasi q.q.o. the corresponding quadratic operator maps from the unite circle into the sphere, from the sphere into the unit circle respectively. Moreover we proved such kind of mappings cannot be positive. We think that such a result will allow to check whether a given mapping is pure channel or not. This finding suggests us to produce a class of non positive mappings. Correspondingly the compliment of this class will have interest in physical works. Since, this set contains all completely positive mappings.

Acknowledgments

The author would like to thank Professors Noboru Watanabe, Masanori Ohya and Si Si for their kind invitation to "Workshop on IDAQP and their applications (3-7 March 2014)". Moreover, the author is grateful to Institute for Mathematical Sciences of the National University of Singapore for the kind hospitality and support.

The author also acknowledges the MOHE grant ERGS13-024-0057 and the Junior Associate scheme of the Abdus Salam International Centre for Theoretical Physics, Trieste, Italy.

References

1. L. Accardi, M. Ohya, Compound channels, transition expectations, and liftings, *Appl. Math. Optimization* **39** (1999) 33–59.
2. S. N. Bernstein, The solution of a mathematical problem related to the theory of heredity, *Uchen. Zapiski NI Kaf. Ukr. Otd. Mat.* 1924, no. 1, 83–115. (Russian)
3. O. Bratteli, D. W. Robertson, *Operator algebras and quantum statistical mechanics*. I, Springer, New York-Heidelberg-Berlin 1979.
4. N. N. Ganikhodzhaev, F. M. Mukhamedov, On quantum quadratic stochastic processes, and some ergodic theorems for such processes, *Uzb. Matem. Zh.* 1997, no. 3, 8–20. (Russian)
5. N. N. Ganikhodzhaev, F. M. Mukhamedov, Ergodic properties of quantum quadratic stochastic processes, *Izv. Math.* **65** (2000) 873–890.
6. U. Franz, A. Skalski, On ergodic properties of convolution operators associated with compact quantum groups, *Colloq. Math.* **113** (2008) 13–23.
7. Yu. I. Lyubich, *Mathematical structures in population genetics*, Springer, Berlin 1992.
8. W. A. Majewski, M. Marciniak, On nonlinear Koopman's construction, *Rep. Math. Phys.* **40** (1997) 501–508.
9. F. M. Mukhamedov, On ergodic properties of discrete quadratic dynamical system on C^*-algebras, *Method of Funct. Anal. and Topology*, **7** (2001) No. 1, 63–75.
10. F. M. Mukhamedov, On decomposition of quantum quadratic stochastic processes into layer-Markov processes defined on von Neumann algebras, *Izvestiya Math.* **68** (2004) 1009–1024.
11. F. Mukhamedov, A. Abduganiev, On Kadison-Schwarz type quantum quadratic operators on $M_2(C)$, *Abst. Appl. Anal.* **2013** (2013), Article ID 278606, 9 p.
12. F. Mukhamedov, A. Abduganiev, On description of bistochastic Kadison-Schwarz operators on $M_2(\mathbb{C})$, *Open Systems & Infor. Dynam.* **17** (2010) 245–253.
13. F. Mukhamedov, A. Abduganiev, On pure quasi-quantum quadratic operators of $M_2(\mathbb{C})$, *Open Systems & Infor. Dynam.* **20** (2013) 1350018.
14. F. Mukhamedov, A. Abduganiev, On bistochastic Kadison-Schwarz operators on $M_2(\mathbb{C})$, *Jour. Phys.: Conf. Ser.* **435** (2013) 012018.
15. F. Mukhamedov, H. Akin, S. Temir, A. Abduganiev, On quantum quadrtic operators on $M_2(\mathbb{C})$ and their dynamics, *Jour. Math. Anal. Appl.* **376** (2011) 641–655.
16. M. A. Nielsen, I. L. Chuang, *Quantum Computation and Quantum Information*, Cambridge Univ. Press, Cambridge, 2000.
17. M. Ohya, D. Petz, *Quantum Entropy and Its Use*, Springer, Berlin 1993.
18. M. Ohya, I. Volovich, *Mathematical foundations of quantum information and computation and its applications to nano- and bio-systems*, Springer, New York, 2011.

19. V. Paulsen, *Completely Bounded Maps and Operator Algebras*, Cambridge University Press, 2003.

20. M. B. Ruskai, S. Szarek, E. Werner, An analysis of completely positive trace-preserving maps on M_2, *Lin. Alg. Appl.* **347** (2002) 159–187.

21. M. Saburov, Quadratic Plus Linear Operators which Preserve Pure States of Quantum Systems: Small Dimensions, *Jour. Phys.: Conf. Ser.* (2014) (in press).

22. E. Stormer, *Positive linear maps of operator algebras*. Springer, 2013.

23. E. Stormer, Positive linear maps of operator algebras, *Acta Math.* **110** (1963) 233–278.

NEW NOISE DEPENDING ON THE SPACE PARAMETER AND THE CONCEPT OF MULTIPLICITY

Si Si

Graduate School of Information Science and Technology
Aichi Prefectural University, Japan
sisi@ist.aichi-pu.ac.jp

Dedicated to Professor Takeyuki Hida's 90th birthday

Our work is in line with the reductionism applied to the study of random complex system, which may be expressed as functionals of a noise obtained by reducing the given phenomena. We understand a noise as a system of idealized elemental random variables formed by independent identically distributed random variables.

We are particularly interested in the noise which is depending on a continuous space parameter, which is an ordered set. We can define a system $E(\lambda)$, $\lambda \in (0, \infty)$ of projections and therefore appeal to the Stone-Hellinger-Hahn type theorem, where the notion of the multiplicity arises as a characteristic of the noise in question.

AMS Subject Classification 2000: 60H40 White Noise Theory

Contents

1. Introduction

We understand noise as a system of idealized elemental random variables depending on a continuous parameter. Standard noises are classified according to the probability distribution and type. We know that there are classes:

i) Gaussian depending on time,

ii) Poisson type depending on time,

iii) Poisson type depending on space.

We recognize these noises by approximation, because a system of continuously many independent identically distributed random variables is not easy to be dealt with. We may understand by approximating such a system by a sequence of countably many independent random variables, that is by digital systems. The limit with reasonable assumptions directs us to the three cases mentioned above. For details, we refer to the paper [4].

The noises i) and ii) are well known, however the case iii) is not so popular. We shall particularly focus our attention to the class iii).

The noise of the type iii)

A Poisson distribution, as is well-known, may arise in the study of the law of small probability, where we recognize that there is a freedom to choose the intensity, denoted by λ, of the limiting Poisson distribution. It is a parameter different from the time t. An interesting interplay between t and λ may be found in a Poisson process, but this is not a topic to be discussed in this chapter.

We are interested in the characterization of probability distributions of various noises. For this purpose, transformation of the probability measure space plays an important role. Some general observation will be given in the next section.

Then, we come to the main topic; namely investigations of the space noise. There, to fix the idea, we forget the time t and will discuss the probabilistic roles of λ, viewed as a *space parameter*. Finally, we shall discuss some connections with decomposition of a Lévy process with a special emphasis on classification of components due to the type of the probability distribution.

2. Invariance of Probability Distributions of Noises

Probability distributions of noises, which are systems of idealized random variables, are introduced on the space of generalized functions. We shall, therefore, use characteristic functionals. Since the noise may be considered to be additive, in a sense, we prefer to take the so-called ψ-functional, which is the logarithm of the characteristic functional. In fact, the ψ-functional is additive for sum of independent random variables, regardless they are ordinary or idealized.

We now recall

i) Gaussian case, that is $\dot{B}(t)$.

The ψ functional is $-\frac{1}{2}\|\xi\|^2$. It involves the L^2-norm, which immediately implies the invariance of Euclidean distance, that is rotations or orthogonal transformations acting on the space of generalized functions on which probability distribution of $\dot{B}(t), t \in R^1$ is introduced.

We know that the infinite dimensional rotation group plays extremely important roles in the white noise analysis.

A trivial note is that we may introduce scale, that is to have $\psi(\xi) = -\frac{\sigma^2}{2}\|\xi\|^2$. We then have a distribution of the same type.

ii) Poisson type depending on time,

We now come to Poisson noises. Most elemental distribution is a single Poisson distribution with intensity λ. Single Poisson noise is not so much interested at present.

iii) Poisson type depending on space parameter.

We wish to introduce the space parameter to a collection of Poisson type distributions which is a family of probability distributions and two of every pair are not of the same type. We can therefore expect to find an important technique fitting for the discussion on the *invariance* depending on the parameter. Actually, the parameter to be introduced is the *intensity* of Poisson distribution. The intensity can be viewed as a space variable, this can be illustrated by the construction of noise of Poisson type, cf Lévy [1] and our recent report [3].

Details will be discussed in the next section.

3. Space Noise

The basic idea of this section is to find how to combine suitably parameterized Poisson distributions (each component is atomic in type) so that the compound distribution satisfies invariance under a certain group acting on the probability measure space.

An atomic distribution with space parameter is a Poisson distribution with intensity λ. We now introduce a slack variable u which denotes the scale (in reality, u does not play an essential role from the viewpoint of the classification according to the type) and plays the role of a "label", as it were, of the intensity. Thus, we have a characteristic function $\varphi(z)$ of the distribution in question expressed in the form:

$$\varphi(z) = \exp[\lambda(e^{izu} - 1)],$$

where $z \in R^1$ and $\lambda, u > 0$.

We now modify the characteristic function. One thing, we change imaginary variable iz to real t since we stick to real. As has been mentioned before, we take the ψ-function. Moreover, we modify it a little to have

$$\psi(t) = \lambda e^{tu}. \tag{3.1}$$

This is acceptable, since ψ-functions of two positive sequences $\{a_n\}$ and $\{ca_n\}$, $c > 0$, are the same up to constant $\log c$. We are interested only in analytic properties. Namely, we take the moment generating function of the sequence $\frac{\lambda^n}{n!}$, $n = 0, 1, 2, \ldots$.

Our discussion now starts with the key function (3.1). Since the variable t runs through R^1, which is one-dimensional Euclidean space, so that we can immediately think of the Affine group $Aff(R^1)$, which is denoted simply by A.

Now define the operators $g(a, b)$, parametrized by (a, b) acting on t-space R^1 such that for $g \in A$

$$g = \{g(a, b) : a, b \in R^1\},$$

determined by

$$g(a, b)t = at + b.$$

In terms of the matrix form, we may write in the form

$$\begin{pmatrix} a & b \\ 0 & 1 \end{pmatrix}$$

$$\psi(g(a, b)t) = \lambda e^{bu} e^{atu}. \tag{3.2}$$

We can see that, by the action $g(a, b)$, a Poisson distribution or its generating function changes in such a way that the intensity λ changes to λe^{bu} and the scale u goes to au.

Thus we have a new function of **different type**, so that we have to have a sum of them. This is the reason why we have to introduce "multiplicity".

The generating function tells us

Proposition 3.1: *By the action of the dilation a the intensity does not change.*

Here is an important remark. The above proposition means that the scale parameter u cannot control the intensity. This fact gives us an important suggestion on the decomposition of a Lévy process. Namely, if the intensity does not change, then the type of the distribution remains within the *same type*. So, changing u does not contribute to the decomposition.

Now consider a representation of the group A on the convex hull $G = \{\psi(g(\cdot,\cdot)t), g \in A\}$ spanned by the generating functions applying the action of the Affine group.

First, fix the parameter λ of the generating function. Take finitely many dilations, say a_j's. Then, we may form

$$\sum_j \psi(g(a_j,0)t)$$

which corresponds to a sum of *independent* Poisson type random variables. Obviously the sum can not be a linear combination, since we do not touch the intensity. Also the sum should be only finite sum because of convergence.

Definition 3.2: Let λ (discrete spectrum) be fixed. The number of the independent variables is called the **multiplicity** of the representation.

In fact, the multiplicity may be called discrete multiplicity corresponding to the point spectrum of the intensity.

It is easy to establish a relationship between the generating functions just obtained and Lévy process with the intensity being fixed but with different jumps as many as the number of the multiplicity.

4. The Intensity Measure

Coming back to the affine group, we now restrict our attention to the shift, i.e. the action of b. Then, we have

$$\psi(g(1,b)t) = \lambda e^{bu} e^{tu}.$$

We may repeat such operations as many times as we wish by changing the amount of the shift and by choosing different u's, we can conclude

Theorem 4.1: *The mapping $\psi(g(1,b)t)$ coming from the shift of t generates a measure, denote it by $dn(\lambda)$ sitting in front of e^{tu} of the ψ function.*

Outline of the proof

1) $g(1, b)$ in the Affine group A acts in such a way that

$$g(1, b) = \lambda e^{bu} e^{tu}.$$

Taking various b in R^1 and apply $g(1, b)$ repeatedly. Then form a *convex hull*

$$K = \{\lambda(\alpha e^{bu} + \beta e^{b'u'}); \alpha, \beta > 0, \alpha + \beta = 1\}.$$

2) The next step is to make the set K wider so that we can find "measures" of the set of intensities. In a sense we wish to have a closure $\bar{K}$.

It is not a good method to introduce a topology in K to have, so we use the classical method due to L. Schwartz distribution (generalized function) method.

Let $K^1(R^1)$ be the Sobolev space of order 1 over R^1. A continuous positive generalized function defined on $K^1(R^1)$ is a measure.

3) Take f in the convex hull K and ξ in $K^1(R^1)$, respectively. Form

$$\mathcal{F} = \{f; \langle f, \xi \rangle = \int f(u)\xi(u)du \geq 0, \int \frac{u^2}{1+u^2}f(u)du < \infty\},$$

where f is continuous.

Obviously $\mathcal{F}$ is a subset of $K^{(-1)}(R^1)$, so that the closure of $\mathcal{F}$ can be defined. Let it be denoted by $\mathcal{F}'$.

Each member of $\mathcal{F}'$ is a "measure" denoted by $dn(u)$.

4) We have a freedom to choose any measure dn to have a function $\int e^{tu} dn(u)$ which can be a generating function, since the sum (integral) preserves the property to be a ψ function because a sum of ψ-function corresponds to a sum of independent random variables.

It is easy to be back to a generating function of a probability distribution by the normalization of the sequence to be a probability distribution.

5) Take any measure dn. By the general theory of measures we have a decomposition

$$dn = dn_c + dn_d,$$

where dn_c is the continuous part and dn_d is the discrete part.

From the discrete part we can choose atoms. For each atom u_d we can find the multiplicity. Precisely

$$\psi(t) = \sum_k n(u_d) e^{tu_d},$$

which is a finite sum, because of the integrability.

The factors e^{bu} with λ generate a measure which may be denoted by $dn(\lambda)$.

The measure is decomposed into two parts: continuous part dn_0 and discrete parts $n_d(\lambda_k)$, which is countable.

It is easily seen that every discrete point λ_k admits the multiplicity (> 1), produced by dilations. While each λ in the continuous part produces infinitesimal random variables of Poisson type, so that further consideration will be done in this direction.

[Note 1] We have to pause for a while to remind the note made when we met a difficulty to manage continuously many independent ordinary random variables. Here a similar situation has happened. In addition, we have to consider the multiplicity by using the projections $E(\lambda)$. This fact will be discussed in a separate paper.

[Note 2] It seems to be better to express the measure dn_0 as $dn_0(\lambda(u))$ or simply by $dn_0(u)$ because the translation of t by b always in the form bu which gives the factor e^{bu}. Such a notation helps us to understand the decomposition of a Lévy process. Namely, we finally have

Theorem 4.2: *The general form of the ψ-function that admits invariance under the Affine transform is expressed in the form*

$$\psi(t) = \int dn_0(u)(e^{tu} - 1) + \sum_k \lambda_k (\sum_j e^{tu_{j,k}}).$$

Here is reminded once again that $dn(u)$ involves a discrete part that produces the multiplicity.

5. Remarks

1) Important Fact

So far we have discussed the characterization of the sequence of positive numbers

$$\left\{ \frac{\lambda^k}{k!}, k = 0, 1, 2, \ldots \right\},$$

by using the action of the affine group $Aff(R^1)$.

Now, let it be back to **probability distribution**. Namely, we put $e^{-\lambda}$ to have

$$\left\{ \frac{\lambda^k}{k!} e^{-\lambda}, k = 0, 1, 2, \ldots \right\}.$$

Having returned to the Poisson distribution, we may revisit a decomposition of a Lévy process from "our" viewpoint (by using Group action).

Since the variable t runs through R^1, it is quite natural to consider the action by $Aff(R^1)$.

Note: If we come to the n-dimensional case, i.e. the case R^n, then we use the group $Aff(R^n)$ which is the skew-product $GL(n, R)$ and R^n:

$$at + b \;\; \rightarrow \;\; At + \mathbf{b},$$

where A is $n \times n$ non-singular matrix and $\mathbf{t}$ is n-dimensional vector. Then, the meaning of the group action gets more clear.

2) Lévy process

Change the variable t of the generating function to iz to have the characteristic function. Then, we understand it corresponds to a decomposition of a Lévy process.

The discrete part n_d of the intensity measure would lead us to a (new) general decomposition of a Lévy process with the **multiplicity**.

Significant remark is the "Decomposition" means that different components should be of different type.

cf. Decomposition of a natural number n to be a product of prime number. This is the meaning of decomposition.

3) Hellinger-Hahn theory

We do not use directly the Stone-Hellinger-Hahn theorem concerning unitary group, but we have used the idea to use the resolution of the identity and the roles of the projection $E(\lambda)$.

4) We have not mentioned on the multiplicity for the case of continuous spectrum $dn(u)$.

If $dn(u)$ satisfies dilation invariant property, then $dn_c(u) = \frac{c}{u^{1+\alpha}}, 0 < \alpha < 2$. Then, certainly the multiplicity is one and we can discuss a topic related to the

$$\psi(t) = \int (e^{tu} - 1) \frac{1}{u^{1+\alpha}} du.$$

Acknowledgement

The author is grateful to "Institute for Mathematical Sciences of the National University of Singapore" for financial support.

References

1. P. Lévy, Pisa Journal paper (1934), 337-366.
2. Si Si, Introduction to Hida distributions. World Sci. Pub. Co, 2012.
3. Si Si, Graded rings of homogeneous chaos generated by polynomials in noises depending on time and space parameters, respectively, Moscow paper 2013.
4. N. Wiener, Nonlinear problems in random theory. The MIT Press, 1958.

A HYSTERESIS EFFECT ON OPTICAL ILLUSION AND NON-KOLMOGOROVIAN PROBABILITY THEORY

Masanari Asano[1], Andrei Khrennikov[2], Masanori Ohya[3,4] and
Yoshiharu Tanaka[3,5]

[1] *Liberal Arts Division, Tokuyama College of Technology*
Gakuendai, Shunan, Yamaguchi, Japan
asano@tokuyama.ac.jp
[2] *International Center for Mathematical Modeling in Physics and Cognitive*
Sciences, Linnaeus University
S-35195, Växjö, Sweden
andrei.khrennikov@lnu.se
[3] *Department of Information Science, Tokyo University of Science*
2641, Yamazaki, Noda, Chiba, Japan
[4] *ohya@rs.noda.tus.ac.jp*
[5] *tanaka@is.noda.tus.ac.jp*

In this study, we discuss a non-Kolmogorovness of the optical illusion in the human visual perception. We show subjects the ambiguous figure of "Schröeder stair", which has two different meanings [1]. We prepare 11 pictures which are inclined by different angles. The tendency to answer "left side is front" depends on the order of showing those pictures. For a mathematical treatment of such a context dependent phenomena, we propose a non-Kolmogorovian probabilistic model which is based on adaptive dynamics.

1. Introduction

In classical probability theory, joint probability of two event systems $P(A,B)$ must hold usual probability law such as a formulae of total probability (FTP), *i.e.* $P(A) = \sum_A P(A,B)$. However, it is known that, in quantum probability theory, FTP is not always satisfied [8, 11, 16]. Also, we can consider quantum probability is a kind of context dependent probability. From this point of view, although quantum mechanics is usually considered as a physics for understanding microscopic phenomena, quantum probability can be applied to non-physical context dependent phenomena, *e.g.* in the fields of biology, cognitive psychology, decision making theory [4, 5, 14, 15, 17]. In this sense, we can say that quantum probability theory is one of non-Kolmogorov probability theories.

We have been discussed the violation of FTP on context dependent phenomena in the various fields: State change of tongue for sweetness [16], Bayesian updating biased by Psychological factor [12], Lactose-glucose interference in *E. coli* growth [10, 13], decision making in two-player game [6, 7, 9].

In this study, we show an optical illusion on the ambiguous figure of Schröder stair [1], and propose its non-Kolmogorov probabilistic model.

First, we introduce the concept of contextual probability, and a definition of joint probability based on adaptive dynamics. Second, we explain Schröder stair and our experiment so as to obtain non-Kolmogorov statistical data. In the last part, we construct a model of visual perception on Shröder stair, based on mathematical frame work of adaptive dynamics.

2. Adaptive Dynamics and Joint Probability

In this section, we reconsider the fundamental concept of probability theory with a simple example, "taste for sweetness".

2.1. *Taste for sweetness: As an example of adaptive phenomena*

Let us consider sweetness of food. If we taste an orange, then some people say that "it is sweet" (1), and the other people say that "it is not sweet" (0). The ratio between 1 and 2 is depend on the orange. Therefore we describe the property of the orange's sweetness as the probability distribution of an random variable which takes 1 or 0.

Here let us consider that one takes other foods, sugar and chocolate. We denote a random variable for chocolate by $C = 1, 0$, and a random variable for sugar by $S = 1, 0$. If one takes either sugar or chocolate, then the probability of answer 1, *i.e.* $P(C = 1)$ or $P(S = 1)$, is very close to 1. However, when one takes chocolate after taking sugar, the tongue will become dull. Therefore most of people cannot taste sweet (maybe no taste). Then $P(C = 1|S = 1)P(S = 1) + P(C = 1|S = 0)P(S = 0)$ is much less than 1. In this case, we have the following inequality.

$$P(C = 1) \neq P(C = 1|S = 1)P(S = 1) + P(C = 1|S = 0)P(S = 0).$$

This inequality is called *violation of formulae of total probability* (FTP), and makes very important discussion for the non-Kolmogorov feature of adaptive (context-dependent) phenomena. The most important thing in the above inequality is that the experimental context of LHS is different from that of RHS. That is, $P(C = 1)$ of LHS is given by the experimental results without taking sugar, but RHS is obtained as the result of both sugar and chocolate. From the point of Kolmogorov's probability theory, the violation is obvious since probability space or measure must be defined within the single experimental context, *e.g.* context of "taking sugar" or "not taking sugar". However, if we discuss the difference of these contexts, the violation of FTP is meaningful. For instance, in order to discuss the change of tongue state, we need to combine these two different contexts. It is necessary to construct such a theory in which the context dependency of the probabilistic phenomena can be discussed.

2.2. *What is adaptive dynamics?*

Adaptive dynamics is a detailed theory to describe the existence (subsistent) by taking various effects surrounding it, and the concept of adaptive dynamics was proposed by Ohya. In the book [11], adaptive dynamics is explained as follows.

> "We find a mathematics to describe a sort of subjective aspects of physical phenomena, for instance, observation, effects of surroundings"

From mathematical point of view, adaptive dynamics is considered that the dynamics of a state or an observable after an instant (say the time t_0) attached to a system of interest is affected by the existence of some other observable and state at that instant. Such dynamics was defined by lifting [3] as follows.

Let $\rho \in S(\mathcal{H})$ and $A \in \mathcal{O}(\mathcal{H})$ be a state and an observable before t_0, and let $\sigma \in S(\mathcal{H} \otimes \mathcal{K})$ and $Q \in \mathcal{O}(\mathcal{H} \otimes \mathcal{K})$ be a state and an observable to give an effect to the state ρ and the observable A. This effect is described by a lifting $\mathcal{E}^*_{\sigma Q}$, so that the state ρ becomes $\mathcal{E}^*_{\sigma Q}\rho$ first, then it will be $\text{tr}_\mathcal{K}\mathcal{E}^*_{\sigma Q}\rho \equiv \rho_{\sigma Q}$. The adaptive dynamics here is the whole process such as

$$Adaptive\ Dynamics: \ \rho \Rightarrow \mathcal{E}^*_{\sigma Q}\rho \Rightarrow \rho_{\sigma Q} = \text{tr}_\mathcal{K}\mathcal{E}^*_{\sigma Q}\rho.$$

That is, what we need is how to construct the lifting for each problem to be studied, that is, we properly construct the lifting $\mathcal{E}^*_{\sigma Q}$ by choosing σ and Q properly. The expectation value of another observable $B \in \mathcal{O}(\mathcal{H})$ in the adaptive state $\rho_{\sigma Q}$ is

$$\mathrm{tr}\rho_{\sigma Q} B = \mathrm{tr}_{\mathcal{H}} \mathrm{tr}_{\mathcal{K}} B \otimes I \mathcal{E}^*_{\sigma Q} \rho.$$

Now suppose that there are two quantum event systems

$$\mathcal{A} = \{a_k \in \mathbb{R}, F_k \in \mathcal{P}(\mathcal{H})\},$$
$$\mathcal{B} = \{b_j \in \mathbb{R}, E_j \in \mathcal{P}(\mathcal{K})\},$$

where we do not assume F_k, E_j are projections, but they satisfy the conditions $\sum_k F_k = I$, $\sum_j E_j = I$ as POVM (positive operator valued measure) corresponding to the partition of a probability space in classical system. Then the "joint-like" probability obtaining a_k and b_j might be given by the *formula*

$$P(a_k, b_j) = \mathrm{tr}\left(E_j \boxdot F_k \mathcal{E}^*_{\sigma Q} \rho\right), \tag{2.1}$$

where $\boxdot$ is a certain operation (relation) between A and B, more generally, one can take a certain operator function $f(E_j, F_k)$ instead of $E_j \boxdot F_k$. If σ, Q are independent from any F_k, E_j and the operation $\boxdot$ is the usual tensor product $\otimes$ so that A and B can be considered in two independent systems or to be commutative, then the above "joint-like" probability becomes the joint probability. However, if this is not the case, *e.g.*, Q is related to A and B, the situation will be more subtle. Therefore, the problem is how to set the operation $\boxdot$ and how to construct the lifting $\mathcal{E}^*_{\sigma Q}$ in order to describe the particular problems associated to systems of interest. Recently we have been discuss this problem in the context dependent systems like bio-systems and psycho-systems mentioned in Introduction.

2.3. *Non-Kolmogorovian probability in adaptive dynamics for sweetness*

Let $|e_1\rangle$ and $|e_2\rangle$ be the orthogonal vectors describing sweet and non-sweet states, respectively. We set the $|e_1\rangle$ and $|e_2\rangle$ as $\binom{1}{0}$ and $\binom{0}{1}$ in the Hilbert space $\mathbb{C}^2$. *The initial (neutral) state of a tongue* is given by a density operator

$$\rho \equiv |x_0\rangle \langle x_0|,$$

with $x_0 = \frac{1}{\sqrt{2}}(|e_1\rangle + |e_2\rangle)$. Here the neutral pure state ρ describes the state of tongue before the experiment, and we start from this state ρ.

When one tastes sugar, the state of tongue will be changed. Such change can be written as the operator on the Hilbert space. The operator corresponding to tasting sugar is represented as

$$S = \begin{pmatrix} \lambda_1 & 0 \\ 0 & \lambda_2 \end{pmatrix},$$

where λ_i are complex numbers satisfying $|\lambda_1|^2 + |\lambda_2|^2 = 1$. This operator can be regarded as the square root of the sugar state σ_S:

$$\sigma_S = |\lambda_1|^2 E_1 + |\lambda_2|^2 E_2, \quad E_1 = |e_1\rangle \langle e_1|, \quad E_2 = |e_2\rangle \langle e_2|.$$

Taking sugar, he will taste that it is sweet with the probability $|\lambda_1|^2$ and non-sweet with the probability $|\lambda_2|^2$, so $|\lambda_1|^2$ should be much higher than $|\lambda_2|^2$ for usual sugar. This comes from the following change of the neutral initial state of a tongue:

$$\rho \to \rho_S = \Lambda_S^*(\rho) \equiv \frac{S^* \rho S}{\operatorname{tr} |S|^2 \rho}. \tag{2.2}$$

This is the state of a tongue after tasting sugar.

The subtle point of the present problem is that, just after tasting sugar, the state of a tongue is neither ρ_S nor ρ. However, for a while the tongue becomes dull to sweetness (and this is the crucial point of our approach for this example), so the tongue state can be written by means of a certain "exchanging" operator $X = \begin{pmatrix} 0 & 1 \\ 1 & 0 \end{pmatrix}$ such that

$$\rho_{SX} = X \rho_S X.$$

Similarly, when one tastes chocolate, the state will be given by

$$\rho_{SXC} = \Lambda_C^*(\rho_{SX}) \equiv \frac{C^* \rho_{SX} C}{\operatorname{tr} |C|^2 \rho_{SX}},$$

where the operator C has the form

$$C = \begin{pmatrix} \mu_1 & 0 \\ 0 & \mu_2 \end{pmatrix}$$

with $|\mu_1|^2 + |\mu_2|^2 = 1$. Common experience tells us that $|\lambda_1|^2 \geq |\mu_1|^2 \geq |\mu_2|^2 \geq |\lambda_2|^2$ and the first two quantities are much larger than the last two quantities.

As can be seen from the preceding consideration, in this example the "adaptive set" $\{\sigma, Q\}$ is the set $\{S, X, C\}$. Now we introduce the following (nonlinear) lifting:

$$\mathcal{E}_{\sigma Q}^*(\rho) = \rho_S \otimes \rho_{SXC} = \Lambda_S^*(\rho) \otimes \Lambda_C^*(X\Lambda_S^*(\rho)X).$$

The corresponding joint probabilities are given by

$$P(S = j, C = k) = \text{tr } E_j \otimes E_k \mathcal{E}^*_{\sigma Q}(\rho).$$

The probability that one tastes sweetness of the chocolate immediately after tasting sugar is

$$P(S = 1, C = 1) + P(S = 2, C = 1) = \frac{|\lambda_2|^2 |\mu_1|^2}{|\lambda_2|^2 |\mu_1|^2 + |\lambda_1|^2 |\mu_2|^2},$$

which is $P(C = 1)$. Note that this probability is much less than

$$P(C = 1) = \text{tr } E_1 \Lambda^*_C(\rho) = |\mu_1|^2,$$

which is the probability of sweetness tasted by the neutral tongue ρ. This means that the usual formula of total probability should be replaced by the adaptive (context dependent) probability law.

3. Optical Illusions in Human Perception Process

In this section, we shortly introduced the phenomena of "optical illusion", and we show the experiment and its results on optical illusion.

A human being does not recognize what comes into our eyes. Our recognition is unconsciously biased. How to bias depends on how it exists in certain surroundings. See the Figure 1. These letters are of the same brightness of color in reality. However we feel the different brightness which is biased by background colors. This is a famous example of optical illusion.

Psychologists explain that this is caused by "perceptual constancy". The perceptual constancy plays a very important role in human perception. When we see a pyramid, the shape of pyramid is different for the angles

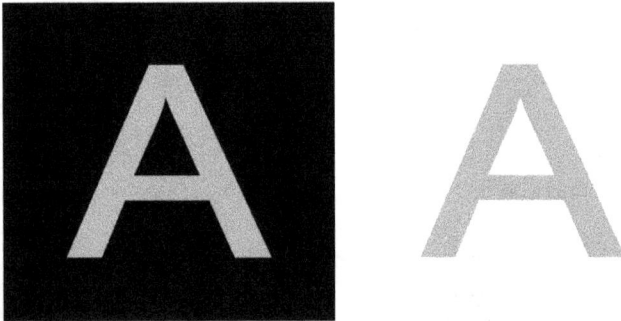

Fig. 1.   Peceptual constancy: Left letter 'A' looks a little brighter than right letter 'A'.

at which we see it. From the side, we can see a triangle, while, from the sky, we can see a square. However, we can recognize this triangle and that square means the same object (a pyramid). It is considered that such stable recognitions of humans are assisted by the perceptual constancy[a].

3.1. *Ambiguous figure*

Ambiguous figure is a figure which has two or more meanings. Rubin's vase is one of most famous ambiguous figures (see Figure 2). The intensity of illusion depends on how it exists in the surroundings [2]. For example, if we change the size of figure, how can we see in the same ambiguous figure? For the tiny figure of Rubin's vase, one can confirm that we have the tendency of seeing "vase" more than "faces".

Fig. 2. Rubins vase: It has two meanings of "two human faces" (white) and "a vase" (black).

3.2. *Depth-inversion for Schröder stair*

Another famous example of ambiguous figure is Schröder stair (see Figure 3).

We can see two different types of stair seen in the Figure 3.

1) The surface of A is front, and the surface of B is back.
2) The surface of B is front, and the surface of A is back.

We can change our recognition consciously or unconsciously. Such phenomena is called "depth inversion". Tendency of recognitions 1) and 2)

---

[a]There is a similar phenomenon that we can feel that snow at night is white (not gray).

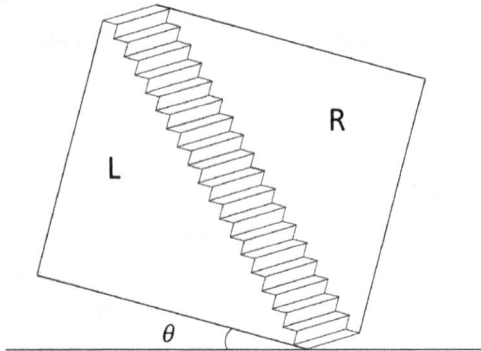

Fig. 3.   Schröder stair: One can see the steps in the center of box (which is leaning).

depends on various features of picture; *e.g.* Relative size of the part A for B, Color (or shadow) in picture, Angle to the horizon, *etc.* We test the tendency with respect to leaning angle.

We show the subjects of experiment the 11 different pictures.

Case A The value of θ is randomly selected from {0, 10, 20, 30, 40, 45, 50, 60, 70, 80, 90}, but each value is selected only once.

Case B The value of θ is regularly changed from 0 to 90.

Case C The value of θ is regularly changed from 90 to 0.

We divided the 151 subjects into three groups: (A) 55 persons, (B) 48 persons, (C) 48 persons. Computer show a picture to a subject. Recognition and decision making. Subjects answer 1) or 2) by means of typing corresponding keys. In case (A), computer decide the order of showing pictures for each subject.

Here we describe the answer for the figure with angle θ as a random variable X_θ such that

$$X_\theta = \left\{ \begin{array}{ll} L & \text{(L is front)} \\ R & \text{(R is front)} \end{array} \right. .$$

Figure 4 shows the probability distribution $P(X_\theta = L)$ with respect to angle θ.

We can see the difference among three cases especially at the middle range of angles. Tendency of human perception is affected by unconscious cognitions for the dynamical change of the angle. We confirm that the probability is decreasing with respect to θ for any cases. However, in the result of case B, the speed of decreasing is faster than that in case A. In

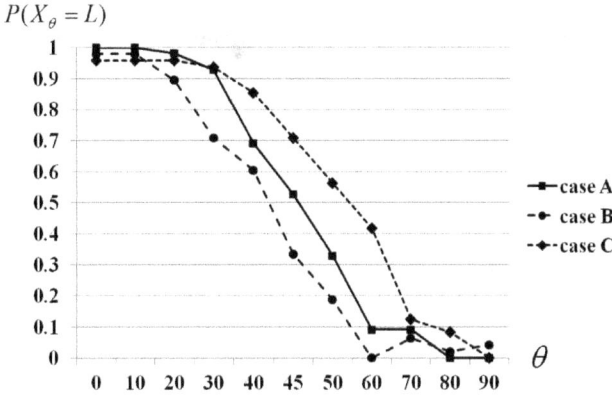

Fig. 4.   Probability distribution $P(X_\theta = L)$ with respect to angle θ.

the case C, we remark that angle θ is decreasing from 90 degrees. Therefore this kind of acceleration effect is similarly seen in case C. These effects may come from perceptual consistency explained in the previous section.

4. Non-Kolmogorovian Model of Optical Illusion

In this section, we show a non-Kolmogorovian model of optical illusion in Schröder stair.

Let us consider a subject making his answer. We hypothesize the following things:

H1 In the subject's brain, there are some imaginary "Agents" which are copies of the subject shown in the picture.

H2 Subject's decision is given by a majority opinion of all the agents.

Conceptually, this type of decision-making is a kind of "self-dialog" which is unconsciously proceeding in the perception process. Further, we add another hypothesis:

H3 Each agent is mathematically expressed as the superposition state of two alternative answers L or R as follows.

$$|x\rangle = \sqrt{\frac{1}{2}}\,|L\rangle + \sqrt{\frac{1}{2}}\,|R\rangle\,.$$

Here $|L\rangle = \begin{pmatrix} 1 \\ 0 \end{pmatrix}$ and $|R\rangle = \begin{pmatrix} 0 \\ 1 \end{pmatrix}$ be the orthogonal vectors describing the answers (i) and (ii), respectively. This state vector $|x\rangle$ describes pre-decision state of the agent, and represents the neutral mind for $|L\rangle$ and $|R\rangle$ before the decision making starts. In the sequel sections, we discuss the more detailed process of decision making, and propose a model based on adaptive dynamics theory.

4.1. *Model of a majority among N agents*

Let $\mathcal{H}$ be the two dimensional Hilbert space $\mathbb{C}^2$. An initial state of a subject's mind is given by

$$\rho \equiv |x\rangle \langle x| \, .$$

When a subject is shown a picture leaning at angle θ, the subject recognizes the lean of the picture. Such a recognition process is a state-change given by the operator

$$M(\theta) = \begin{pmatrix} \cos\theta & 0 \\ 0 & \sin\theta \end{pmatrix} .$$

After the recognition, the state of mind is changed from initial state ρ to an adaptive state

$$\rho_\theta = \Lambda_M^*(\rho) \equiv \frac{M^*\rho M}{\mathrm{tr}\left(|M|^2 \rho\right)} = \begin{pmatrix} \cos^2\theta & \cos\theta\sin\theta \\ \cos\theta\sin\theta & \sin^2\theta \end{pmatrix} .$$

The fluctuation between (i) and (ii) is expressed as the above ρ_θ.

Here, let us recall the hypothesis H1. At the beginning of the process, a subject imagine and create an imaginary agent in the brain, and this agent has own mind which is expressed as the adaptive state ρ_θ. A subject repeatedly create the imaginary agents during an experiment for X_θ. We can consider the adaptive state describing the whole agents as the N-composite state of ρ_θ:

$$\sigma = \underbrace{\rho_\theta \otimes \cdots \otimes \rho_\theta}_{N},$$

where N is the number of the agents which are created by a subject during making the answer (i) or (ii). If a subject takes a short time to answer, then N might be small. After creation of agents (or on the way to create them), the subject can talk with the imaginary agents in the brain, and he/she knows the answer of each agent. Through this dialogue, the subject can know the opinions of all the agents. For the N agents, there are 2^N possible opinions, and one of them is appeared. The subject's answer is determined by reference to the opinions of all the agents.

At the final step of this dialogue, we additionally assumed that the σ is changed into $|L^{\otimes N}\rangle$ or $|R^{\otimes N}\rangle$ since every subject in this experiment can not answer anything except (i) or (ii). Along this assumption, we introduce the observable-adaptive operator Q which describes the process of making a common decision. For example, in the case of $N = 2, 3$ and 4, the operator Q is

$$Q^{(2)} = |LL\rangle \langle LL| + |RR\rangle \langle RR|,$$

$$Q^{(3)} = |LLL\rangle \left(\langle LLL| + \langle LLR| + \langle LRL| + \langle RLL| \right) + |RRR\rangle \left(\langle RRR| + \langle RRL| + \langle RLR| + \langle LRR| \right),$$

$$Q^{(4)} = |LLLL\rangle \left(\langle LLLL| + \langle LLLR| + \langle LLRL| + \langle LRLL| + \langle RLLL| \right) + |RRRR\rangle \left(\langle RRRR| + \langle RRRL| + \langle RRLR| + \langle RLRR| + \langle LRRR| \right).$$

By applying the operator Q to the state of agents σ, the minority opinions of the agents are ignored, and changed to the majority ones. In this sense, we call this decision making as the *majority system* among the N agents.

Here, the lifting $\mathcal{E}_{\sigma Q}^{*} : \mathcal{S}\left(\mathbb{C}^{2}\right) \to \mathcal{S}\left(\mathbb{C}^{2N}\right)$ is defined as

$$\mathcal{E}_{\sigma Q}^{*}(\rho) = \frac{Q\sigma Q^{*}}{\mathrm{tr}\left(|Q|^{2}\sigma\right)} = \frac{Q\left\{\Lambda_{M}^{*}(\rho)\right\}^{\otimes N}Q^{*}}{\mathrm{tr}\left(|Q|^{2}\left\{\Lambda_{M}^{*}(\rho)\right\}^{\otimes N}\right)} \in \mathcal{S}\left(\mathbb{C}^{2N}\right),$$

and the probabilities for X_θ are given with this lifting as

$$P(X_\theta = L) \equiv \mathrm{tr}\left(|L\rangle \langle L|^{\otimes N} \mathcal{E}_{\sigma Q}^{*}(\rho)\right),$$

$$P(X_\theta = R) \equiv \mathrm{tr}\left(|R\rangle \langle R|^{\otimes N} \mathcal{E}_{\sigma Q}^{*}(\rho)\right).$$

In $N = 2, 3$ and 4, the probability $P(X_\theta = L)$ has the following forms:

$$P^{(2)}\left(X_\theta = L\right) = \frac{\cos^4 \theta}{\cos^4 \theta + \sin^4 \theta},$$

$$P^{(3)}\left(X_\theta = L\right) = \frac{\left(\cos^3 \theta + 3\cos^2 \theta \sin \theta\right)^2}{\left(\cos^3 \theta + 3\cos^2 \theta \sin \theta\right)^2 + \left(\sin^3 \theta + 3\sin^2 \theta \cos \theta\right)^2},$$

$$P^{(4)}\left(X_\theta = L\right) = \frac{\left(\cos^4 \theta + 4\cos^3 \theta \sin \theta\right)^2}{\left(\cos^4 \theta + 4\cos^3 \theta \sin \theta\right)^2 + \left(\sin^4 \theta + 4\sin^3 \theta \cos \theta\right)^2}.$$

$P(X_\theta = L)$

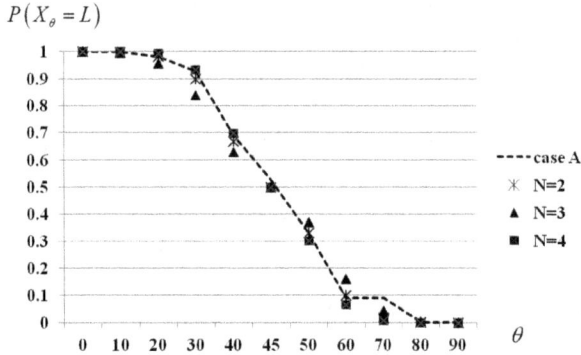

Fig. 5.   Comparison of the values of $P(Xq = L)$ in our model with the experimental data (A).

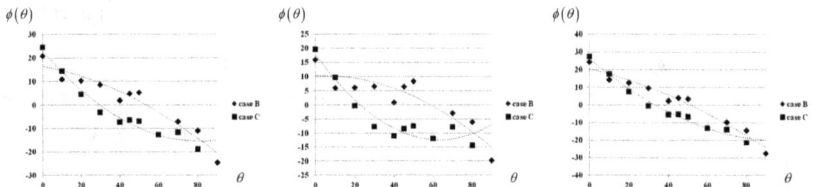

Fig. 6.   Noise of biased phase; (Left) $N = 2$, (Center) $N = 3$, (right) $N = 4$.

We compare these probabilities with experimental data of case (A), see Figure 5. One can find that the probabilities $P(X_\theta = L)$ in $N = 2, 3$ and 4 coincide with experimental data of A.

In the situation (B) or (C), we take another adaptive operator $M(\theta + \phi)$ which angle is shifted by an unknown psychological bias ϕ, instead of $M(\theta)$. This value of ϕ can be also calculated by experimental data. We show it in the Figure 6. The values of ϕ in (B) are higher than those in (C), and especially the difference between $\phi^{(B)}$ and $\phi^{(C)}$ is clearly seen in middle range of angle θ. As increasing N, this difference become smaller.

References

1. H. Schröder, "Ueber eine optische Inversion bei Betrachtung verkehrter, durch optische Vorrichtung entworfener physischer Bilder", Annalen der Physik, 181(10). 298-311 (1858).

2. M. Ohya, "Optical illusions in view of phase transition", Suri-Kagaku (Japanese). 301. 58-61 (1988).

3. L. Accardi, M. Ohya, "Compound Channels, Transition Expectations, and Liftings", Appl. Math. Optim., 39, 33-59 (1999).

4. H. Atmanspacher, T. Filk and H. Römer, "Complementarity in Bistable Perception", Recasting Reality, pp. 135-150, (2009).

5. E. Conte, A. Khrennikov, O. Todarello, A. Federici, L. Mendolicchio and J. Zbilut, "Mental States Follow Quantum Mechanics During Perception and Cognition of Ambiguous Figures", Open Syst. Inf. Dyn. 16, 85 (2009).

6. M. Asano, M. Ohya and A. Khrennikov, "Quantum-Like Model for Decision Making Process in Two Players Game", Founds. of Phys., Springer (2010).

7. M. Asano, M. Ohya, Y. Tanaka, A. Khrennikov and I. Basieva, "On Application of Gorini-Kossakowski-Sudarshan-Lindblad Equation in Cognitive Psychology", Open Syst. Inf. Dyn. 17, 1-15 (2010).

8. A. Khrennikov, "Ubiquitous Quantum Structures: from psychology to finances", Springer (2010).

9. M. Asano, M. Ohya, Y. Tanaka, I. Basieva and A. Khrennikov, "Quantum-like model of brain's functioning: Decision making from decoherence", J. Theor. Biol., 281(1), 56-64 (2011).

10. I. Basieva, A. Khrennikov, M. Ohya and I. Yamato, "Quantum-like interference effect in gene expression: glucose-lactose destructive interference", Syst. Synth. Biol., Springer, 1-10 (2011).

11. M. Ohya and I. Volovich, "Mathematical Foundations of Quantum Information and Computation and Its Applications to Nano-and Bio-systems", Springer-Verlag, (2011).

12. M. Asano, I. Basieva, A. Khrennikov, M. Ohya and Y. Tanaka, "Quantum-like generalization of the Bayesian updating scheme for objective and subjective mental uncertainties", J. Math. Psycho., 56(3), 166-175 (2012).

13. M. Asano, I. Basieva, A. Khrennikov, M. Ohya, Y. Tanaka and I. Yamato, "Quantum-like model of diauxie in Escherichia coli: Operational description of precultivation effect", J. Theor. Biol., 314, 130-137 (2012).

14. E. N. Dzhafarov and J. V. Kujala, "Quantum entanglement and the Issue of selective influences in psychology: An overview", Lecture Notes in Computer Science 7620, 184-195 (2012).

15. J. R. Busemeyer and P. Bruza, "Quantum Models of Cognition and Decision", Cambridge Press (2012).

16. M. Asano, I. Basieva, A. Khrennikov, M. Ohya and I. Yamato, "Non-Kolmogorovian Approach to the Context-Dependent Systems Breaking the Classical Probability Law", Found. Phys., 43 (7), 895-911 (2013).

17. E. Haven and A. Khrennikov, "Quantum Social Science", Cambridge University Press, (2013).

NOTE ON ENTROPY-TYPE COMPLEXITY OF COMMUNICATION PROCESSES

Noboru Watanabe

Department of Information Sciences
Tokyo University of Science
Noda City, Chiba 278-8510, Japan
watanabe@is.noda.tus.ac.jp

Dedicated to Professor Takeyuki Hida

The von Neumann entropy representing the amount of information of the quantum state was introduced around 1932, which was extended by Ohya for C*-systems in 1985 before CNT entropy (1987). The Umegaki relative entropy was defined for σ-finite von Neumann algebras, which was extended by Araki and Uhlmann for general von Neumann algebras and *-algebras, respectively. The Ohya mutual entropy in a complete quantum communication systems was defined in 1983 based on the Ohya compound states instead of the joint probability measures in classical communication systems. It describes the amount of information correctly transmitted through the quantum channel. In this chapter, we briefly explain Ohya's S-mixing entropy and the Ohya mutual entropy for general quantum systems. By using structure equivalent class, we will introduce entropy type functionals based on the quantum entropy theory in order to treat the Gaussian communication process consistently.

1. Introduction

Baker [3, 4] and Skorohod [18] studied Gaussian communication processes mathematically. They computed the classical mutual entropy for the Gaussian measures. In order to discuss the efficiency of information transmission for the communication processes, Shannon's type inequalities by means of the entropies and the mutual entropy with respect to the input state and the communication channels should be satisfied. It was shown in [14] that there exists a simple model for the Gaussian communication process not holding Shannon's type inequalities. To treat the Gaussian communication process consistently, we defined the entropy type functionals in [14] based on the quantum information theory under three conditions (1) linearity condition, (2) trace preserving condition and (3) normality condition. In [21], we studied under two conditions (1) linearity condition and (2) trace preserving condition. The approach developed in [14] was used in the works on classical Gaussian modeling of quantum processes [6].

In quantum information theory, the amount of information of the general quantum state φ is introduced by Ohya's S-mixing entropy $S^S(\varphi)$ [11, 12, 13] and the amount of information correctly transmitted from the input state φ to the output state $\Lambda^*(\varphi)$ through the quantum communication channel Λ^* is given by Ohya's mutual entropy (information) $I(\varphi; \Lambda^*)$ [5, 9, 10, 11, 12, 13, 15]. These entropies satisfy Shannon's type inequalities $0 \leq I(\varphi; \Lambda^*) \leq S(\varphi)$ [9, 10]. Based on the inequalities, we can discuss the efficiency of the information transmission of the communication processes in general quantum systems.

In this chapter, we will improve our previous study [14, 21] under two conditions (1) linearity condition and (2) weak trace preserving condition by using the entropy type functional and the mutual entropy type functional for the Gaussian communication process by means of Ohya's S-mixing entropy and Ohya's mutual entropy for general quantum systems.

2. Gaussian Communication Process

We here briefly explain the Gaussian communication process [3, 4, 14, 15, 18].

Let $\mathcal{B}_k$ be a Berel σ-field generated by all open subsets in a real separable Hilbert space $\mathcal{H}_k$ and let $(\mathcal{H}_k, \mathcal{B}_k)$ $(k = 1, 2)$ be an input and output measurable spaces, respectively. Let $\mathbf{T}(\mathcal{H}_k)_+$ be the set of all positive trace class operators on $\mathcal{H}_k$. We describe by $\mu_k = [m_k, R_k]$ the Gaussian measure μ_k on $\mathcal{H}_k$ with a mean vector $m_k \in \mathcal{H}_k$ and a covariance operator $R_k \in \mathbf{T}(\mathcal{H}_k)_+$ $(k = 1, 2)$. Let $P_G^{(k)}$ be the set of all Gaussian probability measures $\mu_k = [0, R_k]$ on $(\mathcal{H}_k, \mathcal{B}_k)$.

A mapping $\lambda : \mathcal{H}_1 \times \mathcal{B}_2 \to [0, 1]$ is said to be a Gaussian channel from $(\mathcal{H}_1, \mathcal{B}_1)$ to $(\mathcal{H}_2, \mathcal{B}_2)$ if λ satisfies the following two conditions:

(1) For any fixed $x \in \mathcal{H}_1$, $\lambda(x, \bullet)$ is a Gaussian probability measure on $(\mathcal{H}_2, \mathcal{B}_2)$ (i.e., $\lambda(x, \bullet) \in P_G^{(2)}$).
(2) For each fixed $Q \in \mathcal{B}_2$, $\lambda(\bullet, Q)$ is a measurable function on $(\mathcal{H}_1, \mathcal{B}_1)$.

For the Gaussian measure $\mu_0 \in P_G^{(2)}$ representing a noise of the Gaussian communication process, the Gaussian channel λ is given by

$$\lambda(x, Q) = \mu_0(Q^x),$$

where Q^x is

$$Q^x = \{y \in \mathcal{H}_2; \ Ax + y \in Q\}$$

for any $x \in \mathcal{H}_1$ and any $Q \in \mathcal{B}_2$, and A is a linear mapping from $\mathcal{H}_1$ to $\mathcal{H}_2$. For the input Gaussian measure $\mu_1 \in P_G^{(1)}$ and the Gaussian channel λ, the output Gaussian measure $\mu_2 \in P_G^{(2)}$ is obtained by

$$\mu_2(Q) = \Gamma^*(\mu_1)(Q) \equiv \int_{\mathcal{H}_1} \lambda(x, Q)\,\mu_1(dx),$$

where Γ^* is a Gaussian channel from $P_G^{(1)}$ to $P_G^{(2)}$. The compound measure μ_{12} by means of the input measure μ_1 and the output measure μ_2 is given by

$$\mu_{12}(Q_1 \times Q_2) = \int_{Q_1} \lambda(x, Q_2)\,\mu_1(dx)$$

for any $Q_k \in \mathcal{B}_k$ $(k = 1, 2)$. The mutual entropy (Gelfand-Kolmogorov-Yaglom) with respect to the input measure μ_1 and the channel λ is given by

$$I(\mu_1; \lambda) = S(\mu_{12}, \mu_1 \otimes \mu_2),$$

where $S(\mu_{12}, \mu_1 \otimes \mu_2)$ is called a Kullback-Leibler relative entropy with respect to μ_{12} and $\mu_1 \otimes \mu_2$ defined by

$$S(\mu_{12}, \mu_1 \otimes \mu_2) = \begin{cases} \int_{\mathcal{H}_1 \otimes \mathcal{H}_2} \frac{d\mu_{12}}{d\mu_1 \otimes \mu_2} \log \frac{d\mu_{12}}{d\mu_1 \otimes \mu_2} d\mu_1 \otimes \mu_2 & (\mu_{12} << \mu_1 \otimes \mu_2) \\ \infty & (\text{else}) \end{cases}$$

where $\mu_{12} << \mu_1 \otimes \mu_2$ means that μ_{12} is absolutely continuous with respect to $\mu_1 \otimes \mu_2$.

By using the following example, we show a difficulty for the Gaussian communication process found in [14].

Example 2.1: We put $\mathcal{H}_1 = \mathcal{H}_2 = \mathbf{R}^2$. The input and noise Gaussian measures μ_1 and μ_0 are given by $\mu_1 = [0, R_1]$ and $\mu_0 = [0, R_0]$, $R_1, R_0 \in \mathbf{T}(\mathcal{H}_1)_+$.

$$R_1 = \begin{pmatrix} \frac{1}{5} & 0 \\ 0 & \frac{4}{5} \end{pmatrix}, \quad R_0 = \begin{pmatrix} \frac{1}{48} & 0 \\ 0 & \frac{1}{48} \end{pmatrix}$$

We take the linear transformation A by

$$A = \begin{pmatrix} \sqrt{\frac{23}{24}} & 0 \\ 0 & \sqrt{\frac{23}{24}} \end{pmatrix}.$$

Then we obtain the output Gaussian measure $\mu_2 = [0, R_2]$ with

$$R_2 = A R_1 A^* + R_0 = \begin{pmatrix} \frac{17}{80} & 0 \\ 0 & \frac{63}{80} \end{pmatrix}.$$

If we calculate the amount of information of the input Gaussian measure $\mu_1 = [0, R_1]$ by using the definition of the differential entropy, then we have

$$S(\mu_1) = \frac{1}{2} \log \left(\frac{8\pi e}{25} \right) = 0.50265.$$

The classical mutual entropy $I(\mu_1; \lambda)$ of GKY with respect to the input Gaussian measure μ_1 and the Gaussian channel λ is

$$I(\mu_1; \lambda) = \frac{1}{2} \left(\log \left(1 + \frac{46}{5} \right) + \log \left(1 + \frac{184}{5} \right) \right) = 2.9773.$$

Thus the fundamental inequality is not held.

$$S(\mu_1) < I(\mu_1; \lambda).$$

It means that the mutual entropy $I(\mu_1; \lambda)$ with respect to the input Gaussian measure $\mu_1 \in P_G^{(1)}$ and the Gaussian channel λ becomes larger than the differential entropy $S(\mu_1)$ of the input Gaussian measure μ_1. Since the entropy $\tilde{S}(\mu_1)$ by means of all finite partitions of $\mathcal{B}_2$ is infinite, it is impossible to compare the amount of information of the input measures by using $\tilde{S}(\mu_1)$. In order to solve this difficulty, we first studied this problem in [14] based on the quantum information theory. Here we briefly review quantum channel and quantum entropies.

3. Quantum Channels

Let $\mathcal{A}_k$ be von Neumann algebras acting on complex separable Hilbert spaces $\mathcal{H}_k$ $(k = 1, 2)$. We denote the set of all normal states on $\mathcal{A}_k$ by $S(\mathfrak{N}_k)$ $(k = 1, 2)$. Let $(\mathcal{A}_k, S(\mathcal{A}_k))$ $(k = 1, 2)$ be input $(k = 1)$ and output $(k = 2)$ quantum systems, respectively. A mapping from $S(\mathcal{A}_1)$ to $S(\mathcal{A}_2)$ is called a quantum channel Λ^* [1, 5, 8, 12, 13, 15]. Λ^* is called a linear channel if Λ^* satisfies the affine property such as $\Lambda^*(\sum_k \lambda_k \varphi_k) = \sum_k \lambda_k \Lambda^*(\varphi_k)$ for any $\varphi_k \in S(\mathcal{A}_1)$ and any nonnegative number $\lambda_k \in [0, 1]$ with $\sum_k \lambda_k = 1$. For the quantum channel Λ^*, the dual map Λ of Λ^* is defined by

$$\Lambda^*(\varphi)(B) = \varphi(\Lambda(B)), \quad \forall \varphi \in S(\mathcal{A}_1), \quad \forall B \in \mathcal{A}_2.$$

Λ^* is called a completely positive (CP) channel if Λ^* is linear channel and its dual map $\Lambda : \mathcal{A}_2 \to \mathcal{A}_1$ of Λ^* holds a completely positivity. One can describe almost all physical transform of states by using the CP channel [1, 5, 8, 12, 13, 15].

Let $\mathcal{A}'_j$ $(j = 1, 2)$ be two von Neumann algebras acting on complex separable Hilbert spaces representing noise and loss systems, respectively. Quantum communication channel treating the influence of noise and loss is given in [1, 5, 9, 12, 13, 15] by

$$\Lambda^*(\varphi)(B) \equiv \Pi^*(\varphi \otimes \psi)(I \otimes B)$$

for any input state φ in $S(\mathcal{A}_1)$, where ψ in $S\left(\mathcal{A}'_1\right)$ is a given noise state and the mapping Π^* is a CP channel from $S\left(\mathcal{A}_1 \otimes \mathcal{A}'_1\right)$ to $S\left(\mathcal{A}_2 \otimes \mathcal{A}'_2\right)$ by means of the physical properties of the device sending signals [1, 5, 9, 12, 13, 15, 16].

4. Ohya $\mathcal{S}$-Mixing Entropy and Ohya Mutual Entropy for General Quantum Systems

In this section, we review the definitions of the Ohya $\mathcal{S}$-mixing entropy [11, 12, 13] and the Ohya mutual entropy [11, 12, 13] for general quantum systems.

We denote the set of all states on C*-algebra $\mathcal{A}$ by $\mathfrak{S}(\mathcal{A})$ and we represent a unitary semigroup by $\alpha(G)$. A C*-dynamical system is given by a triple $(\mathcal{A}, \mathfrak{S}(\mathcal{A}), \alpha(G))$. $\mathcal{S}$ be a weak* compact convex subset of $\mathfrak{S}(\mathcal{A})$, for example, $\mathcal{S}$ is given by $\mathfrak{S}(\mathcal{A}), I(\alpha), K(\alpha)$. $I(\alpha)$ is the set of all invariant states for α, $K(\alpha)$ is the set of all KMS states. $ex\mathcal{S}$ is the set of all extreme points of $\mathcal{S}$. Every state $\varphi \in \mathcal{S}$ has a maximal measure μ pseudosupported on $ex\mathcal{S}$ such that

$$\varphi = \int_{ex\mathcal{S}} \omega d\mu.$$

The measure μ giving the above decomposition is not unique unless $\mathcal{S}$ is a Choquet simplex. $M_\varphi(\mathcal{S})$ denotes the set of all such measures. $D_\varphi(\mathcal{S})$ is the subset of $M_\varphi(\mathcal{S})$ defined by

$$D_\varphi(\mathcal{S}) = \{M_\varphi(\mathcal{S}); \quad \exists \mu_k \subset \mathbb{R}^+ \text{ and } \{\varphi_k\} \subset exS$$
$$\text{s.t.} \quad \sum_k \mu_k = 1, \quad \mu = \sum_k \mu_k \delta(\varphi_k)\},$$

where $\delta(\varphi)$ is the Dirac measure concentrated on an initial state φ. For a measure $\mu \in D_\varphi(\mathcal{S})$, $H(\mu)$ is given by

$$H(\mu) = -\sum_k \mu_k \log \mu_k.$$

For a C*-dynamical system, the $\mathcal{S}$-mixing entropy [11, 12, 13] of a state $\varphi \in \mathcal{S}$ with respect to $\mathcal{S}$ is defined by

$$S^{\mathcal{S}}(\varphi) \equiv \begin{cases} \inf \{H(\mu); \quad \mu \in D_\varphi(\mathcal{S})\} \text{ if } D_\varphi(\mathcal{S}) \neq \emptyset, \\ +\infty \qquad\qquad\qquad \text{ if } D_\varphi(\mathcal{S}) = \emptyset. \end{cases}$$

For an initial state $\varphi \in \mathcal{S}$ and a channel $\Lambda^* : \mathfrak{S}(\mathcal{A}) \to \mathfrak{S}(\mathcal{B})$, two compound states are introduced by

$$\Phi_\mu^{\mathcal{S}} = \int_{\mathcal{S}} \omega \otimes \Lambda^*\omega \, d\mu,$$
$$\Phi_0 = \varphi \otimes \Lambda^*\varphi.$$

The compound state $\Phi_\mu^{\mathcal{S}}$ expresses the correlation between the input state φ and the output state $\Lambda^*\varphi$. Since the mutual entropy with respect to $\mathcal{S}$ and μ is given by

$$I_\mu^{\mathcal{S}}(\varphi \, ; \, \Lambda^*) = S\left(\Phi_\mu^{\mathcal{S}}, \, \Phi_0\right).$$

For a C*-dynamical system, the Ohya mutual entropy [11, 12, 13] with respect to $\mathcal{S}$ is defined by

$$I^{\mathcal{S}}(\varphi \, ; \Lambda^*) = \sup \left\{I_\mu^{\mathcal{S}}(\varphi \, ; \Lambda^*) \, ; \, \mu \in M_\varphi(\mathcal{S})\right\},$$

which satisfies

$$0 \leq I^{\mathcal{S}}(\varphi \, ; \Lambda^*) \leq S^{\mathcal{S}}(\varphi) \quad \text{and} \quad I^{\mathcal{S}}(\varphi \, ; id) = S^{\mathcal{S}}(\varphi).$$

A example of $I^{\mathcal{S}}(\varphi \, ; \Lambda^*)$ in quantum system denoted by a density operator ρ and a quantum channel Λ^* is the Ohya mutual entropy [9, 10] with respect to the initial state ρ and the quantum channel Λ^* defined by

$$I(\rho; \Lambda^*) \equiv \sup \left\{\sum_n S(\sigma_E, \sigma_0), \rho = \sum_n \lambda_n E_n\right\},$$

where σ_E, σ_0 are the compound states given by $\sigma_E = \sum_n \lambda_n E_n \otimes \Lambda^* E_n$ and $\sigma_0 = \rho \otimes \Lambda^* \rho$ associated with the Schatten-von Neumann (one dimensional spectral) decomposition [17] $\rho = \sum_n \lambda_n E_n$ of the input state ρ, and $S(\cdot,\cdot)$ is the Umegaki's relative entropy [19] denoted by

$$S(\rho,\sigma) \equiv \begin{cases} tr\rho \left(\log \rho - \log \sigma\right) & \text{(when } \overline{ran\rho} \subset \overline{ran\sigma}) \\ \infty & \text{(otherwise)} \end{cases}$$

which was extended to more general quantum systems by Araki and Uhlmann [2, 5, 12, 13, 19, 20]. It also satisfies

$$0 \le I(\rho;\Lambda^*) \le S(\rho) \text{ and } I(\rho; id) = S(\rho),$$

where $S(\rho) \equiv -tr\rho \log \rho$ is von Neumann entropy [7].

5. New Treatment of Gaussian Communication Process

To solve this difficulty, we studied this problem in [14] under the following three conditions:

- **Treatment I**

  (1) Linearity condition (linear approximation)
  (2) Trace preserving condition
  (3) Normality condition

In this chapter, we will improve our previous discussion under the weaker conditions.

We define entropy type functionals for the Gaussian state by using a structure equivalent class, which is defined as follows.

We denote the set of all positive normal functionals on von Neumann algebras $\mathcal{A}_k$ acting on separable Hilbert spaces $\mathcal{H}_k$ ($k = 1, 2$) by $\mathcal{A}_{k*,+}$. For a positive normal functional $\varphi_k \in \mathcal{A}_{k*,+}$, there exists a positive trace class operator $R_k \in \mathbf{T}(\mathcal{H}_k)_+$ (the set of all positive trace class operators on $\mathcal{H}_k$) satisfying

$$\varphi_k(B) = trR_k B, \quad (\forall B \in \mathbf{B}(\mathcal{H}_k)),$$

where $\mathbf{B}(\mathcal{H}_k)$ is the set of all bounded linear operators on $\mathcal{H}_k$. Let Π_k^* be a mapping from $\mathbf{T}(\mathcal{H}_k)_+$ to $\mathcal{A}_{k*,+}$ and $\mathcal{A}'_{k*,+}$ be the set of all positive normal functionals φ_k generated by all operators $R_k \in \mathbf{T}(\mathcal{H}_k)_+$. Then, there exists a bijection Π_k^* from $\mathbf{T}(\mathcal{H}_k)_+$ to $\mathcal{A}'_{k*,+}$.

We study a mapping $\Theta^* : \mathcal{A}'_{1*,+} \to \mathcal{A}'_{2*,+}$ consisted of $\Pi^*_1, \Pi^*_2, \Xi^*_1, \Xi^*_2, \Gamma^*$ defined by the following diagram

$$
\begin{array}{ccc}
\mu_1 \in P^{(1)}_G & \xrightarrow{\;\Gamma^*\;} & \Gamma^*(\mu_1) = \mu_2 \in P^{(2)}_G \\
\updownarrow \Xi^*_1 & & \updownarrow \Xi^*_2 \\
\Xi^*_1(\mu_1) \in \mathbf{T}(\mathcal{H}_1)_+ & & \Xi^*_2(\Gamma^*(\mu_1)) \in \mathbf{T}(\mathcal{H}_2)_+ \\
\updownarrow \Pi^*_1 & & \updownarrow \Pi^*_2 \\
\psi_1 = \Pi^*_1 \circ \Xi^*_1(\mu_1) \in \mathcal{A}'_{1*,+} & \xrightarrow{\;\Theta^*\;} & \Theta^*(\psi_1) = \Pi^*_2 \circ \Xi^*_2 \circ \Gamma^*(\mu_1) \in \mathcal{A}'_{2*,+}
\end{array}
$$

which is denoted by

$$
\begin{aligned}
& \Theta^*(\psi_1)(B) \\
&= \Pi^*_2 \circ \Xi^*_2 \circ \Gamma^* \circ (\Xi^*_1)^{-1} \circ (\Pi^*_1)^{-1}(\psi_1)(B) \\
&= \Pi^*_2 \left(A(\Pi^*_1)^{-1}(\psi_1) A^* + \Xi^*_2(\mu_0) \right)(B), \quad \left(\forall \psi_1 \in \mathcal{A}'_{1*,+}, \forall B \in \mathcal{A}_2 \right)
\end{aligned}
$$

In this chapter, we assume the following two conditions.

- **Treatment II**

  (1) Linearity condition (linear approximation)
  (2) Trace preserving condition : $\Theta^*(\psi)(I) = \psi(I)$ is hold for any $\psi \in \mathcal{A}'_{1*,+}$.

We define a structure equivalent class in the Gaussian communication processes.

Definition 5.1: Structure equivalent of $\mathcal{A}'_{1*,+}$ and $\mathbf{P}^{(1)}_G$

(1) ψ_1 and ψ_2 are structure equivalent (i.e., $\psi_1 \overset{s}{\sim} \psi_2$) if there exists a positive number $\lambda > 0$ such that $\psi_1(I) = \lambda \psi_2(I)$ holds,
(2) $\mu_1 = \left[0, (\Pi^*_1)^{-1}(\psi_1)\right]$ and $\mu_2 = \left[0, (\Pi^*_2)^{-1}(\psi_2)\right]$ are structure equivalent (i.e., $\mu_1 \overset{s}{\sim} \mu_2$) if $\psi_1 \overset{s}{\sim} \psi_2$ is satisfied.

Definition 5.2: Structure equivalent class of $\mathcal{A}'_{k*,+}$ and $\mathbf{P}^{(k)}_G$

(1) $\widetilde{\psi_k} \equiv \left\{ \varphi \in \mathcal{A}'_{k*,+}; \;\; \psi_k \overset{s}{\sim} \varphi \right\}$,
(2) $\widetilde{\mu_k} \equiv \left\{ \nu \in \mathbf{P}^{(k)}_G; \;\; \mu_k \overset{s}{\sim} \nu \right\} \quad (k = 1, 2)$.

By using the above positive constants α and β, one can choose representatives $\alpha \varphi_1$ and $\beta \varphi_1$ of the structure equivalent class of $\mathcal{A}'_{1*,+}$.

Definition 5.3: Quotient sets of $\mathcal{A}'_{k*,+}$ and $\mathbf{P}_G^{(k)}$

(1) $\mathcal{A}'_{k*,+} \big/ \overset{s}{\sim}, \quad \mathcal{A}'_{k*,+} = \bigcup\limits_{\widetilde{\psi_\ell} \in \mathcal{A}'_{k*,+} / \overset{s}{\sim}} \widetilde{\psi_\ell},$

(2) $\mathbf{P}_G^{(k)} \big/ \overset{s}{\sim}, \quad \mathbf{P}_G^{(k)} = \bigcup\limits_{\widetilde{\nu_k} \in \mathbf{P}_G^{(k)} / \overset{s}{\sim}} \widetilde{\nu_k} \quad (k = 1, 2)$

Definition 5.4: Mappings $\widetilde{\Theta}^*$ and $\widetilde{\Gamma}^*$

(1) $\widetilde{\Theta}^*$ is a mapping from $\mathcal{A}'_{1*,+} \big/ \overset{s}{\sim}$ to $\mathcal{A}'_{2*,+} \big/ \overset{s}{\sim}$.

(2) $\widetilde{\Gamma}^*$ is a mapping from $\mathbf{P}_G^{(1)} \big/ \overset{s}{\sim}$ to $\mathbf{P}_G^{(2)} \big/ \overset{s}{\sim}$. We have the following theorems.

Theorem 5.5:

(1) Θ^* is a completely positive map from $\mathcal{A}'_{1*,+}$ to $\mathcal{A}'_{2*,+}$.

(2) $\widetilde{\Theta}^*$ is a completely positive map from $\mathcal{A}'_{1*,+} \big/ \overset{s}{\sim}$ to $\mathcal{A}'_{2*,+} \big/ \overset{s}{\sim}$.

Theorem 5.6: *The Gaussian measure $\bar{\mu} = [0, \Phi_E]$ is a compound state (measure) given by the input measure $\mu_1 = [0, \Xi_1^*(\mu_1)]$ on $\mathcal{H}_1$ and the output measure $\mu_2 = \left[0, (\Pi_2^*)^{-1} \circ \Theta^* \circ \Pi_1^* \circ \Xi_1^*(\mu_1)\right]$ on $\mathcal{H}_2$ in the sense that*

$$\bar{\mu}_1(B) = \bar{\mu}(B \otimes \mathcal{H}_2) \quad \text{for any subspace } B \text{ in } \mathcal{B}_1,$$
$$\bar{\mu}_2\left(B'\right) = \bar{\mu}\left(\mathcal{H}_1 \otimes B'\right) \quad \text{for any subspace } B' \text{ in } \mathcal{B}_2,$$

where $\bar{\mu}_k(B) = \int_B \|\xi\|^2 d\mu_k(\xi)$ for any $B \in \mathcal{B}_k$ and any $\mu_k \in \mathbf{P}_G^{(k)}$ $(k = 1, 2)$.

Let $\mathcal{S}_{1*,+}$ be a weak* compact convex subset of $\mathcal{A}'_{1*,+}$, $\Pi_1^* \circ \Xi_1^*(\mu_1) \in \mathcal{S}_{1*,+} \subset \mathcal{A}'_{1*,+}$ has a maximal measure ν pseudosupported on $ex\mathcal{S}_{1*,+}$ such that

$$\psi_1 \equiv \Pi_1^* \circ \Xi_1^*(\mu_1) = \int_{ex\mathcal{S}_{1*,+}} \omega d\nu,$$
$$\varphi_1 \equiv \frac{\psi_1}{\psi_1(I)} = \frac{\int_{ex\mathcal{S}_{1*,+}} \omega d\nu}{\psi_1(I)}.$$

The structure equivalent for the compound states and the input states are given by

- **Structure Equivalent of the Input State**

  (1) $\mu_1 \overset{s}{\sim} \Gamma^*(\mu_1)$

  (2) $\psi_1 = \Pi_1^* \circ \Xi_1^*(\mu_1) \overset{s}{\sim} \varphi_1 \equiv \frac{\psi_1}{\psi_1(I)}$

Let $M_{\varphi_1}(\mathcal{S}_{1*,+})$ be the set of all such measures belonging to $\mathcal{A}'_{1*,+}$. $D_{\varphi_1}(\mathcal{S}_{1*,+})$ is the subset of $M_{\varphi_1}(\mathcal{S}_{1*,+})$ defined by

$$D_{\varphi_1}(\mathcal{S}_{1*,+}) = \left\{ M_{\varphi_1}(\mathcal{S}_{1*,+}); \quad \exists \nu_k \subset \mathbb{R}^+ \text{ and } \{\psi_k\} \subset ex\mathcal{S}_{1*,+} \right.$$

$$\left. \text{s.t.} \quad \sum_k \nu_k = 1, \quad \nu = \sum_k \nu_k \delta(\psi_k) \right\},$$

where $\delta(\psi)$ is the Dirac measure concentrated on an initial state φ_1. For a measure $\nu \in D_{\varphi_1}(\mathcal{S}_{1*,+})$, $H(\nu)$ is given by

$$H(\nu) = -\sum_k \nu_k \log \nu_k.$$

We first define the entropy type functional of structure equivalent $\tilde{S}_{SE}(\mu_1)$ of the input Gaussian measure $\mu_1 = [0, \Xi_1^*(\mu_1)]$ by using the $\mathcal{S}$-mixing entropy $S^{\mathcal{S}_{1*,+}}(\varphi_1)$ in general quantum system [11, 12, 13] of a state $\varphi_1 \in \mathcal{S}_{1*,+}$ with respect to $\mathcal{S}_{1*,+}$ such as

$$\tilde{S}_{SE}(\mu_1) \equiv S^{\mathcal{S}_{1*,+}}(\varphi_1) \equiv \begin{cases} \inf\{H(\nu); \ \nu \in D_{\varphi_1}(\mathcal{S}_{1*,+})\} \text{ if } D_{\varphi_1}(\mathcal{S}_{1*,+}) \neq \varnothing, \\ +\infty \quad \text{if } D_{\varphi_1}(\mathcal{S}_{1*,+}) = \varnothing. \end{cases}$$

For an initial state $\varphi_1 \in \mathcal{S}_{1*,+}$ and a channel $\Theta^* : \mathcal{A}'_{1*,+} \to \mathcal{A}'_{2*,+}$, two compound states are

$$\Psi_\nu^{\mathcal{S}_{1*,+}} \equiv \int_{\mathcal{S}_{1*,+}} \omega \otimes \Theta^* \omega \, d\nu,$$

$$\Phi_\nu^{\mathcal{S}_{1*,+}} \equiv \frac{\Psi_\nu^{\mathcal{S}_{1*,+}}}{\Psi_\nu^{\mathcal{S}_{1*,+}}(I \otimes I)} = \frac{\int_{\mathcal{S}_{1*,+}} \omega \otimes \Theta^* \omega \, d\nu}{\Psi_\nu^{\mathcal{S}_{1*,+}}(I \otimes I)},$$

$$\Psi_0 \equiv \Pi_1^* \circ \Xi_1^*(\mu_1) \otimes \Theta^* \circ \Pi_1^* \circ \Xi_1^*(\mu_1),$$

$$\Phi_0 \equiv \frac{\Pi_1^* \circ \Xi_1^*(\mu_1) \otimes \Theta^* \circ \Pi_1^* \circ \Xi_1^*(\mu_1)}{\Pi_1^* \circ \Xi_1^*(\mu_1) \otimes \Theta^* \circ \Pi_1^* \circ \Xi_1^*(\mu_1)(I \otimes I)}.$$

The structure equivalent for the compound states and the input states are given by

- **Structure Equivalent of Compound States**

  (1) $\Psi_0 \overset{s}{\sim} \Phi_0$
  (2) $\Psi_\nu^{\mathcal{S}_{1*,+}} \overset{s}{\sim} \Phi_\nu^{\mathcal{S}_{1*,+}}$

The compound state $\Phi_\nu^{\mathcal{S}_{1*,+}}$ expresses the correlation between the input functional $\Pi_1^* \circ \Xi_1^*(\mu_1)$ and the output functional $\Theta^* \circ \Pi_1^* \circ \Xi_1^*(\mu_1)$. Since the mutual entropy with respect to $\mathcal{S}_{1*,+}$ and ν is given by

$$I_\nu^{\mathcal{S}_{1*,+}}(\varphi_1 ; \Theta^*) = S\left(\Phi_\nu^{\mathcal{S}_{1*,+}}, \Phi_0\right).$$

Next we define the mutual entropy type functional of structure equivalent $\tilde{I}_{SE}(\mu_1; \lambda)$ with respect to the input Gaussian measure μ_1 and the Gaussian channel λ associated with $\mathcal{S}_{1*,+}$ such as

$$\tilde{I}_{SE}(\mu_1; \lambda) \equiv \sup\left\{ I_\nu^{\mathcal{S}_{1*,+}}(\varphi_1 \; ; \; \Theta^*) \; ; \; \nu \in M_{\varphi_1}(\mathcal{S}_{1*,+}) \right\}.$$

Based on the treatment of II, we have the following inequalities [22]:

Theorem 5.7: *For any $\mu_1 \in \mathbf{P}_G^{(k)}$ and for some Gaussian channel λ, one obtain the Shannon's type fundamental inequalities as*

$$0 \le \tilde{I}_{SE}(\mu_1; \lambda) \le \tilde{S}_{SE}(\mu_1).$$

In this chapter, we assume the following two conditions.

- **Treatment III**
    (1) Linearity condition (linear approximation)
    (2) Weak trace preserving condition : $\Theta^*(\psi)(I) = \Theta^*(\psi')(I)$ is hold for any $\psi, \psi' \in \mathcal{A}'_{1*,+}$ satisfying $\psi(I) = \psi'(I)$.

We have the following theorems [21].

Theorem 5.8: *If Θ^* satisfies the trace preserving condition, then Θ^* holds the weak trace preserving condition.*

Theorem 5.9: *For $\psi_1 \in \mathcal{A}'_{1*,+}$, if Θ^* satisfies the following inequality*

$$\psi_1(I) < \Theta^*(\psi_1)(I)$$

then there exists a positive number $\alpha > 1$ satisfying the trace preserving condition w.r.t. $\alpha\psi_1$.

Theorem 5.10: *For $\psi_1 \in \mathcal{A}'_{1*,+}$, if Θ^* satisfies the following inequality*

$$\psi_1(I) > \Theta^*(\psi_1)(I)$$

then there exists a positive number $\beta < 1$ satisfying the trace preserving condition w.r.t. $\beta\psi_1$.

$$(\beta\psi_1)(I) = \Theta^*(\beta\psi_1)(I).$$

It means that $\beta\psi_1$ holds the trace preserving condition.

We calculate some examples for treatment III [23]. Now we show Shannon's type inequalities of the entropy functional $\tilde{S}(\mu_1)$ and the mutual entropy functional $\tilde{I}(\mu_1, \lambda)$ for the positive trace class operators under the above treatment III.

Theorem 5.11:

$$\Gamma^* (R_1) = R_2 = AR_1 A^* + R_0$$

If $tr R_1 = 1$ and $A^* A = (1 - tr R_0) I$ then $\Gamma (Q) = A^* QA + (tr R_0 Q) I$ $(\forall Q \in B (\mathcal{H}))$, $\Gamma (I) = I$ and $tr \Gamma^* (R_1) = 1$.

Proof: Since $\Gamma^* (R_1) = R_2 = AR_1 A^* + R_0$ is given,

$$tr R_1 \Gamma (Q) = tr \Gamma^* (R_1) Q$$
$$= tr R_1 (A^* QA + (tr R_0 Q) I)$$

and we have

$$tr \Gamma^* (R_1) = tr (AR_1 A^* + R_0) = 1. \qquad \square$$

Theorem 5.12:

$$\Gamma^* (R_1) = R_2 = AR_1 A^* + R_0$$

If $tr R_1 = \tau > 1$ then $\Gamma (Q) = A^* QA + (tr R_0 Q) I$ $(\forall Q \in B (\mathcal{H}))$ and $tr \Gamma^* (R_1) = \tau + (1 - \tau) tr R_0$ with $\frac{\tau}{\tau - 1} > tr R_0$.

Proof: Since $\Gamma^* (R_1) = R_2 = AR_1 A^* + R_0$ is given, one can obtain

$$tr \frac{R_1}{\tau} \Gamma (Q) = tr \Gamma^* \left(\frac{R_1}{\tau} \right) Q$$
$$= tr \frac{R_1}{\tau} (A^* QA + (tr R_0 Q) I)$$

and we obtain

$$tr R_2 = tr (AR_1 A^* + R_0)$$
$$= \tau + (1 - \tau) tr R_0$$

$$\tau + (1 - \tau) tr R_0 > 0.$$

If $\tau > 1$ then R_0 should satisfies the inequality

$$\frac{\tau}{\tau - 1} > tr R_0 > 0. \qquad \square$$

Theorem 5.13:

$$\Gamma^* (R_1) = R_2 = AR_1 A^* + R_0$$

If $tr R_1 = \tau < 1$ then $\Gamma (Q) = A^* QA + (tr R_0 Q) I$ $(\forall Q \in B (\mathcal{H}))$ and $tr \Gamma^* (R_1) = \tau + (1 - \tau) tr R_0$ with $\frac{\tau}{\tau - 1} > tr R_0$.

Proof: Since $\Gamma^* (R_1) = R_2 = AR_1 A^* + R_0$ is given, one can obtain

$$tr \frac{R_1}{\tau} \Gamma (Q) = tr \Gamma^* \left(\frac{R_1}{\tau} \right) Q$$

$$= tr \frac{R_1}{\tau} (A^* QA + (tr R_0 Q) I)$$

and we obtain

$$tr R_2 = tr (AR_1 A^* + R_0)$$

$$= \tau + (1 - \tau) tr R_0$$

$$\tau + (1 - \tau) tr R_0 > 0$$

is hold for any $(0 <) \tau < 1$. □

Theorem 5.14: *The Gaussian channel Γ^* given by*

$$\Gamma^* (R_1) = R_2 = AR_1 A^* + R_0$$

with $A^* A + (tr R_0) I$ and $\Gamma^* (R_1) > (tr R_0) (\Gamma^* (R_1) - 1)$ holds the weak trace preserving condition.

Proof: For any $R_1, R_1' \in T (\mathcal{H})_+$ with $tr R_1 = tr R_1' = \tau > 1$, since $\tau > (\tau - 1) (tr R_0)$ is hold, we have

$$tr \Gamma^* (R_1) = \tau - (\tau - 1) (tr R_0) = tr \Gamma^* (R_1') > 0.$$

For any $R_1, R_1' \in T (\mathcal{H})_+$ with $tr R_1 = tr R_1' = \tau \leq 1$, one has

$$tr \Gamma^* (R_1) = \tau + (1 - \tau) (tr R_0) = tr \Gamma^* (R_1') > 0.$$ □

Theorem 5.15: *The entropy functional $\widetilde{S} (\mu_1)$ and the mutual entropy functional $\tilde{I} (\mu_1, \lambda)$ with respect to the Gaussian measure $\mu_1 \in P_G^{(1)}$ and the Gaussian channel Γ^* given by*

$$\Gamma^* (R_1) = R_2 = AR_1 A^* + R_0$$

with $A^* A + (tr R_0) I$ and $tr R_1 > (tr R_0) ((tr R_1) - 1)$ satisfies the Shannon's type inequalities such as

$$0 \leq \tilde{I} (\mu_1, \lambda) \leq \tilde{S} (\mu_1).$$

Proof: By using the above theorems, Γ^* holds the weak trace preserving condition. For any $R_1 \in T(\mathcal{H})_+$ with $tr R_1 = \tau \leq 1$, there exists a positive number α such as

$$\alpha = \frac{1}{\tau} \geq 1,$$

one can obtain the equality

$$tr \alpha R_1 = tr \Gamma^* (\alpha R_1) = 1.$$

The entropy functional $\tilde{S}(\mu_1)$ and the mutual entropy functional $\tilde{I}(\mu_1, \lambda)$ satisfies

$$0 \leq \tilde{I}(\mu_1, \lambda) \leq \tilde{S}(\mu_1)$$

where $\tilde{S}(\mu_1)$ and $\tilde{I}(\mu_1, \lambda)$ are obtained by

$$\tilde{S}(\mu_1) = S(\alpha R_1) \quad \text{and} \quad \tilde{I}(\mu_1, \lambda) = I(\alpha R_1, \Gamma^*).$$

For any $R_1 \in T(\mathcal{H})_+$ with $tr R_1 = \tau > (\tau - 1)(tr R_0)$, there exists a positive number β such as

$$\beta = \frac{1}{\tau} < 1,$$

one can obtain the equality

$$tr \beta R_1 = tr \Gamma^* (\beta R_1) = 1.$$

The entropy functional $\tilde{S}(\mu_1)$ and the mutual entropy functional $\tilde{I}(\mu_1, \lambda)$ satisfies

$$0 \leq \tilde{I}(\mu_1, \lambda) \leq \tilde{S}(\mu_1)$$

where $\tilde{S}(\mu_1)$ and $\tilde{I}(\mu_1, \lambda)$ are obtained by

$$\tilde{S}(\mu_1) = S(\beta R_1) \quad \text{and} \quad \tilde{I}(\mu_1, \lambda) = I(\beta R_1, \Gamma^*). \qquad \square$$

In this chapter, we proposed a mathematical treatment III to solve the difficulty of the Gaussian communication process. Using treatment III, we showed that the fundamental inequalities are hold for a simple example of the Gaussian channel. We can use only density operator for the covariance operator of the Gaussian measures under treatment I. In treatment II, we can calculate only the Gaussian channels with trace preserving condition.

References

1. L. Accardi and M. Ohya, Compound channels, transition expectation and liftings, Appl. Math, Optim., **39**, 33-59 (1999).
2. H. Araki, Relative entropy for states of von Neumann algebras, Publ. RIMS Kyoto Univ. **11**, 809-833, (1976).
3. C.R. Baker, Mutual information for Gaussian processes, SIAM J. Appl. Math., **19**, 451-458, (1970).
4. C.R. Baker, Capacity of the Gaussian channel without feedback, Inform. And Control, 37, 70-89 (1978).
5. R.S. Ingarden, A. Kossakowski, and M. Ohya, Information Dynamics and Open Systems, Kluwer, 1997.
6. A. Khrennikov, A classical field theory comeback? The classical field viewpoint on triparticle entanglement, Physica Scripta, **T143**, Article Number: 014013 (2011) DOI: 10.1088/0031-8949/2011/T143/014013
7. J. von Neumann, Die Mathematischen Grundlagen der Quantenmechanik, Springer-Berlin, 1932.
8. M. Ohya, Quantum ergodic channels in operator algebras, J. Math. Appl. **84**, 318–328 (1981).
9. M. Ohya, On compound state and mutual information in quantum information theory, IEEE Trans. Information Theory, **29**, 770-774 (1983).
10. M. Ohya, Note on quantum probability, L. Nuovo Cimento, **38**, 402-404 (1983).
11. M. Ohya, Some aspects of quantum information theory and their applications to irreversible processes, Rep. Math. Phys., **27**, 19-47 (1989).
12. M. Ohya and D. Petz, Quantum Entropy and its Use, Springer, Berlin, 1993.
13. M. Ohya and I. Volovich, Mathematical Foundations of Quantum Information and Computation and Its Applications to Nano- and Bio-systems, Springer, 2011.
14. M. Ohya and N. Watanabe, A new treatment of communication processes with Gaussian channels, Japan Journal on Applied Mathematics, **3**, 197-206 (1986).
15. M. Ohya and N. Watanabe, Foundation of Quantum Communication Theory (in Japanese), Makino Pub. Co., 1998.
16. M. Ohya and N. Watanabe, Construction and analysis of a mathematical model in quantum communication processes, Electronics and Communications in Japan, Part 1, **68**, No. 2, 29-34 (1985).
17. R. Schatten, Norm Ideals of Completely Continuous Operators, Springer–Verlag, 1970.
18. Skorohod, A.V., Integration in Hilbert space, Springer–Verlag, Berlin New York, (1974).
19. H. Umegaki, Conditional expectations in an operator algebra IV (entropy and information), Kodai Math. Sem. Rep., **14**, 59-85 (1962).
20. A. Uhlmann, Relative entropy and the Wigner-Yanase-Dyson-Lieb concavity in interpolation theory, Commun. Math. Phys., **54**, 21–32, (1977).

21. N. Watanabe, On complexity of quantum communication processes, Americal Institute of Physics, **1508**, 334-342, (2013).

22. N. Watanabe, An entropy based treatment of Gaussian communication process for general quantum systems, Open Systems and Information Dynamics, **20**, No. 3, 1340009, 10 pp. (2013).

23. N. Watanabe, Entropy type complexity of quantum processes, Physica Scripta, **2014**, No. T163 (2014).

www.ingramcontent.com/pod-product-compliance
Lightning Source LLC
Chambersburg PA
CBHW050556190326
41458CB00007B/2070